EPA/832-B-00-007
July 2000

Guide To Field Storage of Biosolids

and Other Organic By-Products Used in Agriculture and for Soil Resource Management

U.S. Environmental Protection Agency
Office of Wastewater Management
Washington, DC

U.S. Department of Agriculture
Agricultural Research Service
Beltsville, MD

Notice

Forward

In February 1993, federal standards for the use or disposal of biosolids (40 CFR Part 503) were enacted (Federal Register, 1993). The Part 503 rule addresses land application and beneficial use of biosolids. Included in the rule was a two-year time limit on storage of biosolids for beneficial use. The Part 503 rule did not specifically address management standards and practices for storage of biosolids.

Since the enactment of Part 503, numerous stakeholders, land appliers and biosolids operators have come to understand that there are critical issues associated with successful off-site storage of biosolids (off-site meaning not at the wastewater treatment facility). These issues have not been addressed by code or other guidance documents that are available for reference by biosolids generators and managers, regulatory agencies, or the public.

This guidance document was written to provide a set of consistent Recommended Management Practices for the field storage of biosolids. It identifies three critical control points for managing the system: the wastewater treatment facility, the transportation process, and the field storage site. It provides the elements needed for good site design and operation. This document also stresses the continuing need for partnership and good communication between the biosolids generators and managers responsible for storage and land application to ensure community-friendly operations. The guide targets management practices to address three critical issues: air quality (odors), water quality, and sanitation (pathogens), which have potential environmental, public health and community relations impacts. In the interest of developing a holistic approach to management of organic byproducts, in Chapter 7 there is a discussion of recommendations for storage of organic by-products other than biosolids.

The information in this document represents the collective efforts of a workgroup of professionals with expertise in the generation, processing, transport, field storage, land application, agricultural use, regulation, and public acceptability of biosolids. This group met in June 1997 at Beltsville, MD, to examine the issues and begin framing a set of recommendations for biosolids storage practices. The workgroup continued its effort over a three-year period and has solicited extensive review and comments from a variety of stakeholders and peers. This guide represents the ideas, experience, and knowledge of these scientists and practitioners relative to management of stored biosolids. The key principles for successful biosolids storage as described here are common to numerous storage projects that have been operated successfully in the U.S. It is the desire of the workgroup and contributors to share information and field management techniques that lead to success, and conversely to failure, so that all biosolids managers and states can develop and operate high quality storage programs that support beneficial biosolids use projects

Acknowledgments

This document represents the efforts and contributions of numerous individuals as shown in the Contributors section. Gratitude is expressed to each person involved in the preparation and review of the many drafts leading up to this guide.

The authors are Patricia Millner, Soil Microbial Systems Laboratory, USDA, Agricultural Research Service; Sharon Hogan, Synagro Inc. (formerly Wheelabrator Water Technologies Inc. Bio Grow Division); and John Walker, Municipal Technology Branch, U.S. EPA, Office of Wastewater Management.

For photographs, illustrations, and examples of existing protocols, special appreciation is extended to:

King County Department of Natural Resources
Los Angles County Sanitation District
Maine Department of Environmental Protection
Springfield Regional Wastewater Treatment Plant
St. Croix, Sensory Inc.
USDA-Natural Resource Conservation Service

We also gratefully acknowledge the assistance of Shannon Garland in editing and layout of the final publication, Connie Kunzler in editing workgroup reports and drafts, and Dorothy Talmud for typing the manuscript.

Contributors

Mary Jo Aiello, Bureau of Pretreatment &
Residuals, N.J. Dept. Environmental
Protection
Trenton, NJ 08625

Robert Bastian
U.S. EPA, Office of Water
Washington, D.C. 20460

Douglas Borgatti
Springfield Water & Sewer Commission
Springfield, MA 01101

J.Scott Carr
Black & Veatch
Kansas City, MO 64114

Gene de Michele
Water Environment Federation
Alexandria, VA 22314-1994

Elliott Epstein
E & A Environmental
Canton, MA 02021

Ervin Faulman
Biocheck Labs
Toledo, OH 43606

Jeffrey Faust, BioGro Div.
Wheelabrator Water Technologies, Inc.
Millersville, MD 21108

Robert A. Gillette
Carollo Engineers
Sacramento, CA 95833

Wes Gregory, President
Waste Stream Environmental
Jordan, NY 13080

Sam Hadeed
National Biosolids Partnership
Alexandria, VA 22314

George Hall
Metropolitan Water Reclamation
District of Greater Chicago
Willow Springs, Ill. 60480

Larry Hentz
Post, Buckley, Schue, Jernigan
Bowie, MD 20716

Penny Hill
Los Angeles County Sanitation District
Whittier, CA 90607

Sharon Hogan, BioGro Div.
Wheelabrator Water Technologies, Inc.
Baltimore, MD 21224

John Hoff
City of Columbus, Composting Facility
Lockbourne, OH 43137

Lee Jacobs
Dept. Crop & Soil Sciences
Michigan State University
East Lansing, MI 48824_1325

Carolyn Jenkins, New England Interstate
Water Pollution Control Commission
Wilmington, MA 01887
Raymond J. Kearney
Hyperion Treatment Div.
Dept. Public Works, City of Los Angeles
Playa Del Rey, CA 90293

Greg Kester
Dept. of Natural Resources
Madison, WI

Mark King
Dept. Environmental Protection-Maine
Augusta, ME 04333

Mark E. Lang
The Sear_Brown Group
Rochester, NY 14623

Richard Litz
Waste Stream Environmental/Earth Blends
Weedsport, NY 13166

Terry Logan
N-Viro International Inc.
Toledo, OH 43606

Pete Machno
National Biosolids Partnership
Seattle, WA

Patricia D. Millner
USDA-Agricultural Research Service
Beltsville, MD 20705

Richard G. Mills
Massachusetts Water Resources Authority
Boston, MA 02129

J. Patrick Nicholson
N-Viro International Inc.
Toledo, Oh 43606

Bob Odette
Synagro
Advance, NC 27006

Randy R. Ott, County of Onondaga
Dept. Drainage and Sanitation
Syracuse, NY 13204

Anthony Pilawski,
Bureau of Pretreatment & Residuals
N.J. Dept. Environmental Protection
Trenton, NJ 08625

Frank Post , AMSCO
P.O. Box 1770
Clemmons, NC 27012

Ben Price
The Merriwood Corp.
Fallbrook CA 92088

Mark Ronayne, City of Portland
Bureau of Environmental Services
Portland OR 97203

Sally Rowland
NY-Dept. Environmental Protection
Albany, NY 12233

Alan B. Rubin (4304)
U.S. EPA, Office of Water
Washington, D.C. 20460

A. Robert Rubin
North Carolina Cooperative Extension
Service, North Carolina State University
Raleigh, NC 27695

L. Douglas Saylor, District Mining Operations
PA-Dept. Environmental Protection,
Hawk Run, PA 16840

John Sendera
Calumet Water Reclamation District
Chicago, IL 60628

Jim Smith
U.S. EPA, NRMRL- TSD
Cincinnati, OH 45268

Bob Southworth (retired)
U.S. EPA , Office of Water
Washington, D.C. 20460

John Stapelton
Waste Stream Environmental/Earth Blends
Weedsport, NY 13166

Steve Stark
Metropolitan Council Environmental Service
St. Paul, MN 55101

Dan Sullivan
Department: Crop & Soil Science
Oregon State Univ.
Corvallis, OR

Michael Switzenbaum
Civil and Environmental Engineering Dept.
Univ. Mass.
Amherst, MA 01003

Joel Thompson
WSSC
Laurel, MD

WilliamToffey
Philadelphia Water Department
Philadelphia, PA 19107

John Walker (4204)
U.S. EPA, Office of Water
Washington, DC 20460

David Wanucha
Synagro
Advance, NC 27006-9801

Neil Webster
Webster Environmental Associates
Pewee Valley, KY 40056

Clyde Wilber
Greeley & Hansen
Upper Marlboro, MD 20772

William Yanko
Los Angeles County Sanitation District
Whittier, CA 90601

Contents

Chapter 1

Introduction

Successful biosolids land application programs should have provisions to deal with daily biosolids production in the event biosolids cannot be land applied immediately. This contingency planning generally includes storage as well as other back-up options, such as landfill disposal, incineration or alternative treatment and use, including composting, heat drying and advanced alkaline stabilization.

Storage is necessary during inclement weather when land application sites are not accessible and during winter months when land application to snow covered and frozen soil is prohibited or restricted. Storage also may be needed to accommodate seasonal restrictions on land availability due to crop rotations or equipment availability. For small generators, storage allows accumulation of enough material to efficiently complete land application in a single spreading operation. Well-planned and managed storage options not only provide operational flexibility at the treatment facility, but they also can improve the agronomic, environmental, and public acceptance aspects of biosolids use.

The focus of this document is on management practices for field storage of biosolids prior to land application, as distinguished from land application and spreading. The document stresses recommended management practices for three critical control points: the WWTP, the transportation system, and the field storage site. The term <u>critical control point,</u> as used in this document, means a location, event or process point at which specific monitoring and responsive management practices should be applied. If these points are controlled, the objectives and goals of a responsible and community-friendly practice can be achieved. Equally important is the continuing need for partnership and good communication among biosolids generators, storage site managers and land appliers.

The term <u>field storage</u> as used in this document refers to temporary or seasonal storage. Storage operations involve an area of land or facilities constructed to hold biosolids until material is land applied on designated and approved sites. More permanently constructed storage facilities can involve state or locally permitted areas of land or facilities used to store biosolids. The permissible time limits for field storage vary by state and local jurisdiction. They are usually located at or near the land application site, and are managed so that biosolids come and go on a relatively short cycle, based on weather conditions, crop rotations, and land or equipment availability. Alternatively, storage sites are used to accumulate enough material to conduct an efficient spreading operation. Some of the terminology frequently used to describe is shown in the box below. The terminology, as well as associated prescribed limits on field storage, can vary from state to state. Definitions of these and other specialized terms that appear elsewhere in this document (as individual bold typeface), and abbreviations can be found in the Glossary (Appendix F).

It is very clear to all biosolids generators, transporters, storers, land appliers, and local officials that malodors are the greatest reason for public concern about storage sites. Much of this guidance document seeks to provide information and strategies useful in minimizing odor problems.

Frequently Used Field Storage Terminology

Staging Field placement of biosolids at the time of delivery in such manner as to facilitate land-application the same day or within a few days; may also involve transfer of biosolids from transport vehicles to equipment for immediate land application.

Stockpiling Holding of biosolids at an active field site long enough to accumulate sufficient material to complete the field application efficiently.

Field storage Temporary or seasonal storage area, usually located at the application site, which holds biosolids destined for beneficial use on designated fields. State regulations may or may not distinguish between staging, stockpiling or field storage. Time limits for storage range from 24 hours to two years, depending on the jurisdiction in which it is located.

Storage Facilities An area of land or constructed facilities committed to hold biosolids until the material may be land applied at on- or off-site locations. This facility may be used to store any given batch of biosolids for up to two years. However, most are managed so that biosolids come and go on a shorter cycle based on weather conditions, crop rotations and land availability, equipment availability, or to accumulate sufficient material for efficient spreading operations.

The types of biosolids discussed in this document include Class A and Class B (classes indicate the degree of pathogen reduction, see Chapter 4). These biosolids are produced by treatment processes that generate liquid, dewatered, heat dried, air-dried, composted, digested, or alkaline stabilized materials. The type and intensity of the treatments varies and this impacts the properties of the biosolids that are placed in storage. Thus, each site should be designed to adequately handle the types of materials expected. The operations management plan should be matched to the properties of the designed site and the type of biosolids being stored.

Management for Storage

This section explains some of the general principles underlying the management of biosolids in storage situations. Biosolids managers should keep these concepts in mind as they assess their storage needs and options and develop a management plan suited to their unique situation.

Critical Control Points (Key Management Areas)

Even with a wide variety of biosolids and the numerous types of field situations that are encountered throughout the U. S., all field storage operations can be broken down into three areas that are critical for good management: the biosolids generating facility, transportation, and the actual field storage site (see box below). Activities in each of these areas are critical to the overall success of biosolids storage operations. For instance, the level of treatment and post-treatment handling at the generating facility may affect the odor characteristics once biosolids reach the field site.

CRITICAL CONTROL POINT 1: *WWTP*
CRITICAL CONTROL POINT 2: *TRANSPORTATION PROCESS*
CRITICAL CONTROL POINT 3: *FIELD STORAGE SITE*

This guide provides detailed descriptions of the practices recommended for management of these areas as well as explanations of their importance relative to odors, water quality, pathogens and community acceptance. Biosolids managers are encouraged to carefully analyze their own particular situations and select the most feasible combination of practices for their unique situation from this guide.

Table 1.1 highlights the main issues and some of the self-monitoring activities and control options involved in each of these management areas. Complete descriptions of these practices are provided in subsequent chapters.

Variables Related to Intensity of Management

There are five variables that affect the level or intensity of management required for successful field storage of biosolids.

1. <u>Stability of Biosolids</u>: Material that is less well stabilized generally has a greater potential to generate unacceptable levels of odorous compounds.

2. <u>Water Content of Biosolids</u>: Liquid and some semi-sold material require pumping equipment and constructed storage facilities.

3. <u>Length of storage period</u>: Longer storage periods increase the potential for exposure to wet or hot weather and a resumption of microbial decomposition leading to the generation of odorous compounds.

4. <u>Volume of stored material</u>: Management requirements in terms of site design, operation and the potential scale of odor or water quality impacts may increase with the volume of material stored.

5. <u>Climate and weather conditions</u>: Warm humid weather or wet conditions generally increase management requirements as compared to storage during dry or cold conditions.

The preceding variables are interrelated and therefore exceptions to particular points may occur when mitigated by other variables. For instance, a large volume of a well-stabilized biosolids may be less management intensive in terms of preventing nuisance odors, than the storage of a small volume of a less well-stabilized material. Figure 1.1 provides a schematic to illustrate several of these interrelated factors. Throughout this guide, the diagram will be used to highlight the importance that the interaction of these factors has on the overall success of biosolids storage operations.

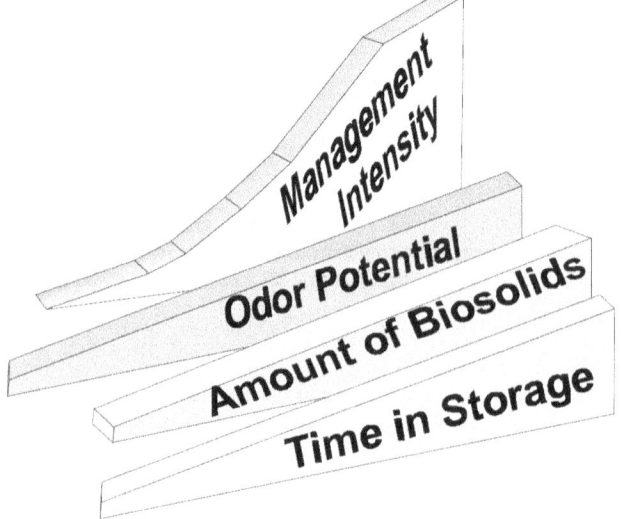

Table 1-1. Overview of Management Control Points for Field Stored Materials*

Issues	Self-Monitoring Checklist	Control Options
1 Biosolids Generating Facility		
• Odors and aesthetics • Consistency of biosolids • Biosolids treatment	• Assess biosolids to determine: • 503 treatment criteria for pathogens and VAR • Degree of stability and odor potential includes factors such as volatile solids content; lime, polymer and iron usage, and pH • Physical consistency • Ratio of primary to secondary • Cleanliness of equipment • Time of retention after treatment	• Generator, storer and land applier communicate about status of biosolids treatment or problems • Reduce post-treatment retention • Have options to divert unacceptable loads • Reevaluate treatment and handling practices to address chronic issues • Provide further treatment • Provide vehicle cleaning station
2. Transportation		
• Odors and aesthetics • Traffic and safety	• Proper equipment in compliance with State and Federal Transportation Regulations • Regular inspections and maintenance of vehicles and equipment • Suitable haul routes • Vehicles and equipment kept clean	• Train drivers • Plan/inspect haul routes, minimize time in transport • Emergency spill plan and supplies in place • Maintain and clean trucks and equipment regularly
3. Field Storage Site		
• Odors and aesthetics • Water quality and environmental protection • Safety and health protection	• Proper site location & suitability • Proper design of field storage or constructed facility - run-on and run-off controls - accumulated water control - buffers • Biosolids quality vs length & amount in storage • Operations and maintenance plan • Odor prevention and mitigation plan • Spill control and response plan • Safety plan	• Regular self inspections of site and operations • Consistent implementation of management plans • Self monitoring of biosolids quality and condition • Revision of management plans when necessary - change amount or length of storage - implement odor control and mitigation measures - implement additional structural or site management practices • Remove stored biosolids when atmospheric conditions are conducive to low odor impacts on neighbors

*See Chapter 5 for recommendations for specific facility/storage options.

Need for Partnerships

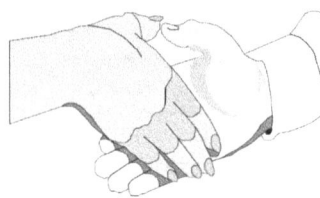

It is recognized by experienced biosolids management teams that partnership and good communication between the biosolids generator and the biosolids manager responsible for storage and land application is essential to optimizing the management of biosolids destined for storage. Successful storage programs require coordination of management activities at the generating facility, in transit and at the storage site.

Likewise, good communication links are necessary between the biosolids manager and the biosolids users, local governments, and citizens of communities where biosolids storage activities are located. Chapter 6 discusses methods to establish and enhance communication links between biosolids managers and communities.

The absence of such partnerships has often resulted in odors or other problems with subsequent unfavorable community acceptance, political, or economic consequences. Land appliers overwhelmed with community relations problems may be forced to cease land application and seek alternative management options. These are typically more costly to consumers than field storage and land application, or result in lost economic and environmental benefits to farmers, landowners, and diversion of biosolids to non-beneficial uses, such as land filling or incineration.

Figure 1-1. Successful biosolids storage programs begin with good communications between biosolids generators and haulers. Pro-active communication and interaction among generators, field operation managers, and neighboring communities facilitate the success of beneficial use programs.

Biosolids Storage Areas

- Storage areas are located on flat, easily accessible terrain
- Storage meets specific field requirements
- Applications of stored biosolids occur from Mar - Nov
- Stabilized (digested) dewatered biosolids have sticky, peat-like consistency

- Over-winter storage occurs from Nov - Mar
- Storage is necessary to meet production, distribution, and market demands for biosolids
- Frozen ground restricts applications

- Storage piles after 5 months of winter storage
- These piles experienced precipitation which exceeded 10-year average by 35%

- Piles remained stable with no movement via overland flow or leaching
- These observations are supported by soil sampling results

Figure 1-2. Good site selection and field management practices ensure that field stockpiles can be used during several seasons (Courtesy of King County, WA, Dept. Natural Resources in cooperation with Boulder Park, Inc)

Chapter 2

Odors
Introduction

Malodors are the single most important cause of public dissatisfaction with biosolids or other organics recycling and utilization projects. Thus, odor management is a high priority. Experience and practice have demonstrated that biosolids and other organic by-products, such as animal manure, landscape trimmings, and food processing residuals, can be handled and processed without release of excessive malodorous compounds. However, if any of these materials, including biosolids, are poorly managed, then objectionable odors may develop during storage, and public acceptance of such a project will erode.

This section provides basic information about odor and describes the practices and rationale for various approaches that are used to minimize odor during storage. A variety of options are available, and it will be necessary for the biosolids manager to determine which ones provide the flexibility needed to accommodate the range of situations in their program. There is no "one size fits all" solution. Chapter 5 has details on odor prevention and mitigation practices.

What Is Odor?

The malodorous compounds (odorants) associated with biosolids, manures, and other organic materials are the volatile emissions generated from the chemical and microbial decomposition of organic nutrients. When inhaled, these odorants interact with the odor sensing apparatus (olfactory system) and the person perceives odor.

Individual sensitivity to the quality and intensity of an odorant can vary significantly, and this variability accounts for the difference in sensory and physical responses experienced by individuals who inhale the same amounts and types of compounds. This distinction between "odor", which is a *sensation*, and "odorant", which is a volatile chemical compound, is important for everyone who deals with the odor issue to recognize. When odorants are emitted into the air, individuals may or may not perceive an odor. With biosolids, three conditions are necessary to create malodorous conditions.

BASIC CONDITIONS ASSOCIATED WITH MALODOROUS SITUATIONS

1. **EMISSION:**   **Presence of an odorous volatile chemical (odorant)**

2. **TRANSPORT:**   **Topographic and atmospheric conditions conducive to transport of the odorant with minimal dilution**

3. **PERCEPTION:** **People are present and they perceive odor**

When people perceive what they regard as unacceptable amounts or types of odor, odorous emissions can become an "odor problem".

Primary Biosolids Odorants

The odorous compounds generated, and most often detectable, at significant levels during biosolids treatment, storage, and use are ammonia, amines and reduced sulfur-containing compounds (for detailed descriptions of these compounds see Appendix B). Amines can be produced in easily detectable quantities during high temperature processes. Amines include: methylamine, ethylamine, trimethylamine, and diethylamine. Amines often accompany ammonia emissions, and if chlorine is used chloramines may be released. The sulfur compounds include compounds such as hydrogen sulfide, dimethyl sulfide, dimethyl disulfide, and methyl mercaptan. The potential for these compounds to be annoying is based in part on their individual and combined quantity, intensity, pervasiveness, and character (see Appendix B for details and definitions).

Amines and reduced-sulfur compounds may be detectable and perceived at greater distances from a storage facility than ammonia because they are more persistent (pervasive), intense, and have very low odor detection thresholds (i.e., people can detect just a few parts per billion in fresh air). Although ammonia is usually the primary odor associated with limed or alkaline stabilized biosolids; it has an intense odor that can often mask other odors, such as reduced-sulfur compounds. However, because the detection threshold for ammonia is much greater than that of many of the reduced sulfur compounds (i.e., it takes more ammonia in air to be detectable than it does sulfur compounds), the odors of reduced-sulfur or amine compounds are more likely than that of ammonia to be detected at distances from the site where ammonia is no longer above its odor threshold concentration.

Odor Management: A Partnership Effort

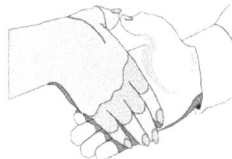 There is no doubt that untreated wastewater solids have inherently undesirable odor qualities. However, many current treatment processes have the capacity to produce biosolids that are minimally odorous. Despite this, occasional malodorous batches can occur, and thus biosolids generators, storers and land appliers should make provisions to handle these appropriately. These provisions rely on close communication and working linkages among the biosolids management partners (i.e., generator, transporter, storer, and applier). Good management of each process technology and a cooperative effort among the biosolids management partners to ensure proper transportation, handling, and storage of the materials can minimize the potential for unacceptable odor concentrations at storage sites.

Minimizing Odor during Storage

- **Stabilize biosolids at WWTP as much as possible**

- **Avoid use of polymers that lead to malodor**

- **Maintain proper pH during treatment**

- **Meet Part 503 Vector Attraction Reduction**

- **Locate storage at remote sites**

- **Minimize duration of storage**

- **Assess meteorological conditions before loading and unloading**

- **Ensure good housekeeping**

Factors Affecting Ultimate Odor Potential at Critical Control Point 1: The WWTP

The following section addresses *Critical Control Point 1* issues. Specific situations and conditions associated with biosolids preparation at the WWTP are described along with their relation to storage and especially odors. When an odor situation cannot be averted, management of the emissions and quick response through mitigation practices are required to avoid creating nuisance odor situations. At the WWTP, which is *Critical Control Point 1*, this coordination includes:

- Assessing the stability of the biosolids before they leave the WWTP

- Having contingency plans to provide remedial treatment, or diversion of unacceptably odorous material to suitable land application or disposal sites.

- Notifying the storer and land applier of any changes in mixing (primary or secondary solids), polymer or other additives, pH, moisture content, or stability.

Decisions relative to odor control are a series of trade-offs involving higher degrees of treatment at the WWTP versus the intensity of management at the off-site storage locations. Ensuring that the odor of biosolids leaving the WWTP is minimized is a key consideration, since it is more difficult to treat an odor problem that originated at the WWTP once the biosolids are placed at the storage site. In all cases, the temporary measures invoked to deal with unexpected and unanticipated events that lead to odors must be considered only as such. Persistent problems will require an examination of the treatment and handling processes to develop a better management approach***.

Stability

The success of the various solids treatment technologies with regard to malodor reduction depends largely on the degree of stabilization achieved in the biosolids before it leaves the treatment facility and the preservation of stability until used. Wastewater treatment technologies differ in their capacity to stabilize biosolids.

The potential for odorous emissions depends partly on the extent to which organic matter and nutrients are present in forms that microbes readily use. Stabilization processes may either: 1) decrease the level of volatile organic compounds and the availability of nutrients to reduce the potential for microbial generation of odors; or 2) change the physical or chemical characteristics of the biosolids in a way that inhibits microbial growth. Table 2-1 lists seven commonly used stabilization and/or processing methods. Odor issues associated with each method and/or process are shown along with appropriate corresponding prevention or remediation approaches.

Table 2-1. Prevention and management of odorous emissions associated with biosolids stabilization or processing methods.

Stabilization and Processing Methods	Potential Causes of Odorous Emissions	Long Term Potential Solutions	Short term Temporary Solutions
Anaerobic Digestion	'Sour', overloaded or thermophilic digester; volatilization of fatty acids and sulphur-compounds	Optimize digester; don't overload	Apply topical lime to stored biosolids
Aerobic Digestion	Low solids retention time; High organic loading, Poor aeration	Increase retention time and aeration; Lower organic load	
Drying Beds	Incomplete digestion of biosolids being dried	Optimize digestion	
Compost	Poor mixing of bulking agent; poor aeration; Improperly operating biofilters.	Mix better; adjust mix ratio and aeration rate; improve biofilter function	Aerate more effectively; remix; re-compost.
Alkaline Stabilization	Addition of insufficient alkaline material so pH drops below 9, microbial decomposition may occur with generation of odorous compounds. Check compatibility of polymer with high pH and other additives, e.g. $FeCl_3$.	Increase pH Provide finer mesh grade of alkaline material and mix better to avoid inadequate contact with biosolids	Check pH; apply topical lime
Thermal Conditioning & Drying	High temperature volatilization of fatty acids and sulfur-compounds	Use secondary treatment biosolids; primary solids are less stable and more odorous when heated.	Apply topical lime if biosolids are still liquid

Digested and Composted Biosolids
Properly digested and/or composted biosolids meet stabilization and vector attraction reduction requirements because these extended treatments reduce pathogens and decompose volatile solids (i.e., the organic matter which serves as food for microbes). When such materials are placed in proper storage, they typically do not contain enough readily available nutrients to support a large, rapid growth of microbes that might generate odorous volatiles.

Alkaline and Chlorine Treated Biosolids
Chemical stabilization processes act to inhibit the growth of microorganisms, rather than to decompose the organic matter in the biosolids. Addition of alkaline materials, such as lime, elevates the pH to levels that suppress microbial activity and kill pathogens. As long as the pH remains high in stored materials, no new potential odorants will be produced. Small residual levels of reduced sulfur or amine compounds, which were generated prior to and not released during stabilization, may be present. One of the sulfur products of concern, hydrogen sulfide, is converted into a non-soluble (non-volatile) form at high pH. Raising the pH will liberate ammonia and amines, especially at the time of treatment. For the ammonia, this is unlikely to result in objectionable

off-site impacts because ammonia is not a persistent odorant. However, amines can be persistent and are more likely detected off-site once ammonia has dissipated and thus stopped masking the amines. In addition, when stored, alkaline stabilized biosolids quickly develop a dry crust, which seals the pile and prevents significant volatilization. Disturbing piles during load-out operations exposes fresh surfaces to the atmosphere and increases the potential for volatilization of trapped residual odorous compounds. Hence, avoid load-out during air temperature inversions and periods of low turbulence, since pervasive odorants will more likely be detected under such conditions.

Drying Beds and Thermal Drying etc.

Heat and/or desiccation are the primary means of pathogen reduction in thermal treatment or drying; these methods also halt microbial decomposition of organic materials. They do not appreciably reduce organic matter during the relatively short time periods in which drying is conducted, and thus require appropriate management during storage to prevent significant resumption of microbial decomposition and release of odorants.

Other Odor Prevention Considerations

The type of treatment and stabilization processes used at a WWTP are primary factors influencing the type and level of odors which may be potentially generated by a particular biosolids. Other factors at the wastewater treatment plant that may affect the odor potential of biosolids include:

Other Important Factors at the Wastewater Treatment Plant that Affect the Odor Potential of Biosolids

- Periodic changes in influent characteristics (e.g. fish wastes, textile wastes and other wastewaters with high odor characteristics)

- Type of polymer used and its susceptibility to decomposition and release of intense and pervasive odorants such as amines when biosolids are heated or treated with strong alkaline materials

- Blending of primary and secondary biosolids which may create anaerobic conditions or stimulate a resumption of microbial decomposition

- Completeness of blending and mixing, and quality of products used for stabilization (i.e. type of lime and granule size)

- Effectiveness and consistency of Vector Attraction Reduction Method, use of Part 503 VAR options 1-8 (treatment at WWTP) vs. VAR options 9-10 (at land application site)

- Handling, storage time, and storage method when stabilized biosolids are held at the WWTP prior to transport (e.g. anaerobic conditions developing in enclosed holding tanks when material is held for several days during hot weather).

Vector Attraction Reduction
Stabilization treatment may include processes at the WWTP to reduce the attraction of vectors to biosolids as outlined in the Part 503 rule. The effectiveness and consistency of these treatments may also help to minimize odor potential. Odor is typically less of a problem for biosolids that fully meet one of the first eight Part 503 VAR options (See Appendix C). However, sometimes it is necessary to store materials that will meet VAR by options 9 or 10 (injection or soil incorporation). In such cases, increased management intensity (e.g. storage for short periods of time, storage during cold weather, storage at remote locations, etc.), and self-monitoring for unacceptable odor levels may be needed to prevent nuisance odor conditions.

Factors Affecting Ultimate Odor Potential at Critical Control Point 2: The Transportation Process

The process of transporting biosolids from the generating facility to the field storage site may impede traffic, be unsightly and can potentially emit nuisance odors into the community. The transportation process (referred to as Critical Control Point 2 in this document) must be properly managed as to minimize these problems, including the public's exposure to biosolids odors. One way to reduce public exposure to odors is to choose a hauling route that avoids densely populated residential areas. The fewer residences located along a hauling route, the less likely the general public will be annoyed by the traffic and biosolids odors. Making sure that the trucks used to haul biosolids are clean and well maintained is another effective way to keep road surfaces clean and control odors during biosolids transport. Trucks should be cleaned before leaving the generating facility and after the biosolids have been deposited on the field storage site. These steps are important because odor concerns are exacerbated by increased road congestion, and by biosolids adhering to trucks and roadways.

Factors Affecting Ultimate Odor Potential at Critical Control Point 3: The Field Storage Site

In most cases, biosolids produced at WWTPs with well-operated stabilization processes can be stored off-site without creating odor nuisances. However, if certain conditions occur while material is in storage, the potential for odorous emissions (sulfur- or amine-containing compounds or ammonia) will increase.

Specific Storage Site Conditions that Contribute to Generation of Odorants • Meteorological conditions • Distance to sensitive receptors (i.e. housing developments) • pH drops below 9 in lime stabilized biosolids • Anaerobic or deficient oxygen conditions within the biosolids • Storage of primary biosolids with waste activated (digested) biosolids • Rewetting of dried material • Ponded water in contact with stored biosolids • Prolonged storage of inadequately stabilized biosolids • Inadequate handling methods • Deficient housekeeping and spill control

Several of the specific site conditions will be discussed later in this chapter or in Chapter 5.

Meteorological Conditions

Meteorological conditions such as wind speed and direction, cloud conditions, relative humidity, and temperature, all of which can change with the season, day to day, and even with the time of day. Warm temperatures and high humidity increase the potential for odor nuisances, while cold, dry conditions reduce the potential for nuisance complaints.

Most odors from a biosolids storage site are area source rather than point source, ground level emissions. Under moderate atmospheric stability (e.g., partly sunny, wind speeds 8-12 mph, moderate turbulence), on flat terrain area source odorants undergo fairly rapid dilution as the distance from the source increases. As such, concentrations of odorants will likely not be objectionable to neighbors, if the biosolids are reasonably well stabilized. Conversely, pervasive odorants from poorly stabilized batches can be detected at considerable distances from the source. Rough terrain, valleys, and other topographical features can increase the complexity of airflow patterns. Odor dispersion analysis can help site managers schedule operations to avoid high odor concentrations from developing at sensitive downwind locations.

Odorants emitted from ground-level sources will remain most concentrated during periods of high atmospheric stability, such as occur with air temperature inversions and low wind speeds at night and very early morning. This means that odor complaints may be higher during non-business hours. Dispersion is enhanced once the sun has warmed the soil surface. For permanent constructed facilities, a basic wind dispersion analysis of the site, including seasonal and annual prevailing wind direction, and typical meteorological conditions for the area will help site operators plan activities so as to minimize odorous emission impacts downwind.

Planning and Monitoring

Whether biosolids are stored in field stockpiles or constructed facilities, odor prevention and mitigation measures need to be part of the operational plans. Also, a complaint response plan to promptly and effectively investigate and respond to local odor concerns or complaints (see Chapter 5 for details on odor prevention and mitigation) also needs to be included. The plan should include provisions for diversion of odorous batches to alternate sites that are remote or other disposal options. In the sections that follow, a notably greater level of effort is required to control odors for constructed facilities than for field stockpiles.

Field Stockpiles

Persons responsible for storage of biosolids should realize that odor is a perceptual, subjective, and frequently emotional issue. In most storage scenarios (particularly small-scale field stockpiles), sophisticated analysis of odorous compounds is not necessary to resolve community odor issues. What is necessary, is a well thought out and implemented odor prevention and mitigation plan designed to be sensitive to local odor concerns. Such a plan should include provisions for prompt response, investigation and follow-up if odor complaints are received (See Chapter 5 for details).

Constructed Facilities

Odor prevention and minimization plans are generally needed for large, longer-term facilities. These plans may need to rely in part on some type of monitoring to determine the extent of odor, and the effectiveness of the procedures used to mitigate odors.

Because sensitivity to the quality and intensity of an odor can vary significantly among individuals, specialized approaches are needed to evaluate the impact of odorous compounds. Odor and Odor Event Characterization Monitoring is a simple, direct approach that relies on odor detection reporting and wind direction recordkeeping. This approach might be considered in place of complex chemical quantification and identification. In this approach, a set of odor characters (descriptors) is identified for use by site operators conducting routine odor inspections and by citizens reporting odor detection events. Examples of odor characters include: sharp pungent (ammonia), unpleasant putrid (amyl mercaptan), pungent suffocating (chlorine), skunk-like (crotyl mercaptan), fishy (trimethylamine), decayed cabbage (dimethyl disulfide), etc. (see Table B-2 for additional descriptors). The odor characters selected should cover the range of odors potentially emitted from a biosolids site, as well as other nearby operations that may also emit distinctive odors.

To the extent possible, descriptors should be identified that can serve as markers for emissions from biosolids. In this way, biosolids managers can focus corrective actions when appropriate. This also will a means to distinguish biosolids odors from those generated by other types of odor emitting facilities in the same area as the storage site, to the extent that they are present.

In order to use the odor descriptors correctly, site managers, personnel, and odor investigators would be trained in the proper use of the odor character descriptors. They would also be trained to recognize field conditions acceptable for selected odor measurements, i.e., intensity and descriptors, and key areas and times for inspection. A simple written report (see Appendix B for example) of odor inspections/investigations can be used to document performance at a site. On-site inspections coupled with use of an immediate odor response plan, can aid in reducing the potential for odor complaints. In some cases, an 'odor hotline' for citizen complaints can be useful. If complaints are received, the storage facility operator is able to promptly dispatch personnel to follow-up with the caller and initiate an investigation and problem remediation.

Recent advances in odor science, detection/recognition threshold determination, and measurement of odor annoyance have helped to reduce the subjective nature of odor evaluation for biosolids (see Appendix B for details). Measurement, Identification, and Monitoring in response to persistent odor problems that need remediation may involve characterizing the source and type of odorants. This requires sophisticated collection, identification, measurement, and evaluation of gases in air samples. Subsequently, the human sensitivity to these odorants is evaluated in terms of their perceived intensity, pervasiveness, and/or annoyance in the impact zone. This also requires specialized measurement equipment and techniques and may benefit from atmospheric dispersion modeling. Obviously, this relatively complex approach to odor assessment would be used in only those biosolids storage situations in which less intensive approaches had failed to lead to remediation, or if the size, nature and storage capacity of the facility required it.

Length of Storage and Changes in Biosolids Characteristics

Preventing the resumption of microbial activity in biosolids is a primary means of controlling odors at storage sites. Microbial decomposition is likely to occur if the pH of lime stabilized biosolids drops below nine; if anaerobic or deficient oxygen conditions occur within the biosolids (free O_2 concentration less than 15 percent); if primary biosolids are mixed and stored with waste activated (digested) biosolids; or if dried material are rewetted. Ensuring that the materials brought to the facility are thoroughly stabilized and minimizing the length of time materials are kept in storage are two major tools to achieve this goal. In some cases, microbial activity can be halted or controlled by on-site remedial actions such as the addition of lime to lagoons or top-dressing stockpiles with lime slurries, or covering of dried materials.

Accumulated Water and Site Management

Precipitation or runoff that accumulates in contact with biosolids will pick up nutrients and organic matter that promote rapid blooms of microorganisms that rapidly deplete dissolved oxygen levels and lead to anoxic or septic conditions and the generation of significant odors. Proper design and operation of the facility as described in Chapter 5 is key to preventing this problem. Establishing good housekeeping procedures and keeping the storage area, equipment and trucks clean and free of standing water is another component of avoiding odor generation. Likewise, conducting handling operations in a clean and efficient manner that minimizes the time materials are disturbed will help limit odor.

References

Borgatti, D., G.A. Romano, T.J. Rabbitt, and T.J. Acquaro. 1997. 1996 Odor Control Program for the Springfield Regional WWTP. New England WEA Annual Conf., 26-29 January 1997, Boston, MA.

Haug, R.T. 1990. An essay on the elements of odor management. Biocycle. 30(10): 60-67.

Hentz, L.H., C. M. Murray, J.L. Thompson, L.L. Gasner, and J.B. Dunson Jr. 1992. Odor control research at the Montgomery Country Regional Composting Facility. Water Environ. Res 64(1): 13-18.

Lue-Hing, C., D.R. Zenz, and R. Kuchenrither. 1992. Municipal Sludge Management-Processing, Utilization, Disposal, Water Qual. Mgmt. Libr (Vol 4),Technomic Publ Co, Inc. Lancaster, PA

McGinley, C.M., D.L. McGinley, and K.J. McGinley. 1995. "Odor School"-Curriculum Development for Training Odor Investigators, pp. 121-127. *In* Air Water Mgmt. Assoc. Intl. Specialty Conf Proc. *Odors and Environmental Air.* Bloomington, MN, 13-15 September 1995.

McGinley, M.A. 1995. Quantifying Public Perception of Odors in a Community St. Croix Sensory, Inc. Stillwater, MN.

Rosenfeld, P. 1999. Characterization, Quantification, and Control of Odor Emissions from Biosolids Application to Forest Soil. Ph.D. Dissertation. University of Washington, Seattle, WA.

Schiffman, S. S.; Walker, J. M.; Dalton, P.; Lorig, T. S.; Raymer, J. H.; Shusterman, D.; Williams, C. M. Potential health effects of odor from animal operations, wastewater treatment, and recycling of byproducts. Journal of Agromedicine, 7: 2000, in press. Available from Haworth Document Delivery Service 1-800-342-9678 or getinfo@haworthpressinc.com.

Smith, J. E. and J. B. Farrell. 1992. Vector Attraction Reduction Issues Associated with the Part 503 Regulations and Supplemental Guidance, U.S. EPA, Center for Environmental Research Information, Cincinnati, OH

Switzenbaum, M.S., L..H. Moss, E. Epstein, A.B. Pincince. 1997. Defining Biosolids Stability: A Basis for Public and Regulatory Acceptance. WERF Project 94-REM-1 Final Report, Water Environ.Res. Foundat., Alexandria, VA.

Vesilind, P.A., G.C. Hartman, and E.T. Skene. 1986. Sludge Management and Disposal for the Practicing Engineer, Lewis Publishers, Inc., Chelsea, MI

Walker, J.M. 1993. Control of Composting Odors, pp. 185-218. In H.A.J. Hoitink and H.M. Keener (eds.), Science and Engineering of Composting Renaissance Publ., Worthington, Ohio.

Walker, J.M. 1991. Fundamentals of odor control. Biocycle 30(9): 50-55.

Wilber, C. and C. Murray. 1990. Odor source evaluation. BioCycle 31(3): 68-72.

Wilber, C. (ed.) 2000. Operations and Design at the Wastewater Treatment Plant to Control Ultimate Recycling and Disposal Odors of Biosolids. USEPA sponsored project.

Wilby, F.V. 1969. Variation in recognition odor threshold of a panel. J.Air Pollut. Contr. Assoc. 19(2):96-100.

Yonkers Joint WWTP. 1997. Process compatibility testing D. Odor. In Specifications for Furnishing and Delivering Liquid Emulsion type polymer (40-50 percent active) for Centrigure dewatering of sludge. Yonkers Joint SSTP, Ludlow Dock, South Yonkers, NY.

Chapter 3

Water Quality

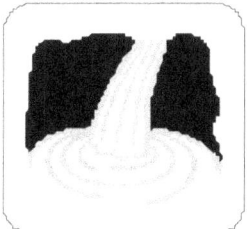

Introduction

This chapter describes the types of water quality impacts potentially attributable to specific nutrients and pollutants in stored biosolids and other organic materials. In addition, key concepts in construction and management of storage systems that are known to work well in preventing water quality impacts from biosolids storage are discussed and related specifically to storage management practices recommended in Chapter 5 (Critical Control Point 3).

Water Quality Issues

Measurements of the following constituents of organic and inorganic materials stored on and/or applied to soil are customarily made to assess their potential impact on water quality. Table 3-1 summarizes these components relative to biosolids storage and their potential impacts on water quality:

- **Nutrients**

- **Organic matter**

- **Pathogens**

- **Metals**

Assessment of the presence of constituents such as nutrients, organic matter, pathogens or metals is the first step in developing effective water quality protection practices for stored materials. The second step is to examine the possible ways of transport. Constituents can only have an impact on water quality if significant amounts of the material reach surface or ground water. Good storage design and use of appropriate management practices effectively block potential transport pathways.

Movement of constituents is driven primarily by:

1. precipitation events

2. runoff and erosion of soluble and particulate components (including nutrients, organic matter, and pathogens to surface waters)

3. leaching to ground water of soluble nutrients and compounds.

In addition, wind erosion can contribute to loss of dry or composted material under arid, windy conditions that may also impact water quality.

Nutrients, Organic Matter, and Impacts on Surface Water

The content and form of nitrogen (N) and phosphorus (P), which must be taken into consideration, in specific biosolids, vary depending on wastewater sources and treatment processes. Like all organic residuals, biosolids contain significant amounts of N and P. Proper storage conserves these nutrients until crops can use them during the growing season. Good management of stored organic residuals is needed to prevent excess amounts of organic or inorganic N from entering surface or ground water.

Runoff of nutrients can contribute to eutrophication of surface water, which may impair its use for fisheries, recreation, industry and drinking water source. Nitrogen is the primary contributor to eutrophication in brackish and saline waters (e.g., estuaries), and to some extent in freshwater systems. However, P concentration is usually the controlling eutrophication factor in freshwater. Both nitrate and ammonia are water soluble, and thus, are transported in leachate and runoff. Organically bound N does not interact in the environment until it is mineralized into water soluble nitrate. Ammonia can be toxic to fish.

Excess nutrients and organic matter in surface water can increase the growth of undesirable algae and aquatic weeds. The carbon and nutrients in organic matter serve as food for bacteria, thus adding organic matter and nutrients to water can directly increase BOD, deplete dissolved oxygen levels in water, and accelerate eutrophication. The amount of oxygen needed to decompose the organic matter that is suspended in the water is called the Biochemical Oxygen Demand (BOD). Low oxygen levels resulting from high BOD stress fish, shellfish and other aquatic invertebrates. In a worst-case scenario, such as a direct spill of material from a storage facility into a waterway, heavy organic (BOD) and ammonia-N loadings could deplete oxygen levels rapidly and lead to septic conditions and fish kills.

The majority of P binds to mineral and organic particles and is therefore subject to runoff and erosion rather than leaching, except under conditions of very sandy soils with low P binding capacity. Eroded particulates also serve as a physical substrate to convey adsorbed P, metals and other potential pollutants in runoff.

Nutrients and Groundwater

The main concern with groundwater impacts of longer-term stockpiles (organic or inorganic) is the potential for leaching of soluble nitrate-N, which can impact local wells or eventually discharge to surface waters and contribute to eutrophication. Such situations have occurred in agricultural regions of the U.S. where excessive amounts of inorganic fertilizer or animal manures have been applied over several years. The high nitrate levels in wells have resulted in some cases of methemoglobinemia in susceptible infants. This rare condition reduces the blood's ability to carry oxygen efficiently, hence the condition's other name "blue baby syndrome." Elevated nitrate in water can have the same effect on immature horses and pigs and can cause abortions in cattle. Water management practices at storage sites must be adequate to protect against such impacts. Phosphorus is not a drinking water concern, because it is not a health concern for humans or animals as nitrate is, and it binds to iron and soil minerals and has low water solubility.

Pathogens

In the U.S., biosolids that are stored prior to land application must have been treated to meet USEPA Part 503 Class A or Class B pathogen density limits. The requirements for these types of biosolids can also include restricted access to field sites (Class B) to protect humans and animals from infection that might potentially result from direct contact with biosolids. Protection of water sources from contamination by residual pathogens or parasites in Class B biosolids can be accomplished through proper site selection, buffers and management practices as described in Chapter 5.

In general, soil is an effective barrier to the movement of pathogens via leachate into groundwater. Both organic matter and clay minerals in soil physically filter, adsorb, and immobilize microorganisms, including protozoan cysts, and parasitic worm ova. However, sandy soils are typically very porous and cannot adsorb or immobilize microbes as well as clay and loam soils containing organic matter, thus they are not as effective retardants to the movement of pathogens. Soils in general are subject to a range of physical, chemical, and biological conditions that destroy pathogens such as: extremes of wetness and dryness; temperature variations; and attack by natural soil microbes.

Metals and Synthetic Organic Chemicals

Like other residuals, biosolids may also contain measurable levels of metals and synthetic organic chemicals. In terms of organic and inorganic residuals, the same management practices that effectively isolate nutrients from surface and groundwater resources during storage are equally effective in containing any metals or synthetic organic chemicals. The potential for water quality impacts from metals or synthetic organic chemicals present in biosolids are minimal from the outset because of their inherently low levels. Biosolids suitable for land application must meet stringent quality standards for metal concentrations under Part 503 regulations.

With the widespread implementation of industrial pretreatment programs, biosolids used in land application increasingly comply with the most conservative of Part 503 metal standards. In addition, metals in biosolids are bound strongly with other biosolids constituents and, are not highly water soluble, hence they cannot leach into ground water. According to a recent review by the National Research Council (NRC), toxic organic compounds typically are not found in biosolids in significant levels. This is primarily attributable to effective industrial pretreatment programs and to the destruction or volatilization of organics during the treatment process. The NRC report also noted that "PCBs and detergents are the only classes of synthetic organic compounds that occur in biosolids at concentrations above levels found in conventional irrigation water or soil additives". PCBs bind to particulates and are relatively water insoluble and so are not susceptible to leaching. In addition, the low PCB levels in biosolids continue to decline due to enactment of a ban on production and use of PCBs in the United States. Detergent compounds including surfactants and binders have been found in biosolids in relatively high concentrations (0.5 - 4.0 g/kg dry weight), however they bind to biosolids organic matter, rapidly biodegrade, and do not readily leach.

Table 3-1. Potential Ground and Surface Water Quality Impacts Resulting from Improper Management of Water at Storage Sites

Biosolids Constituent	Potential Water Quality Impacts	Behavior, Transport Mechanism, and Mitigating Factors
Nitrogen	Eutrophication Human/Livestock/ Poultry health effects	Nitrate-N, Nitrite-N, and Ammonium-N are water soluble and can move in runoff or leachate
Phosphorus	Eutrophication	Predominately particulate-bound transported by erosive surface runoff
Organic Matter	Depletes oxygen levels in water	Soluble and particulate-bound movement of organic matter in surface runoff.
Particulates	Siltation or turbidity. Carrier for other pollutants	Mass transport in surface runoff.
Pathogens	Transmission of viable disease-causing bacteria, viruses or parasites	Insignificant levels in Class A biosolids, potentially present in Class B materials. Physical transport in sediment, runoff, and leachate from Class B biosolids is possible.
Regulated Metals	Toxic effects	Not very water soluble; reduced through pretreatment programs and Part 503 limits.
Toxic Organic Chemicals	Toxic effects	Reduced through industrial pretreatment programs and WWTP processes.

Management Approaches

This section summarizes the key storage design and management elements that address the water quality issues identified in Table 3-1. Water quality protection practices are based on three key concepts:

Protecting Water Quality during Storage of Biosolids:

- Keep clean runoff clean by minimizing contact with stored biosolids.

- Properly manage water that comes into contact with stored biosolids.

- Prevent movement of the biosolids into water resources

Keep Clean Water Clean

Minimizing the amount of water that comes into contact with stored biosolids is the first step in keeping nutrients and pollutants out of water resources. Practices used under various storage scenarios to achieve this include:

- Proper site selection to avoid run-on, flooding, or high water tables that intercept stored biosolids (see Chapter 5 also).

- Installation of upslope diversions to channel runoff away from a field stockpile or constructed storage facility (see Appendix C also).

- Containment of biosolids in enclosed structures or tanks.

Manage Water that Contacts Biosolids

Any significant precipitation or up-slope runoff that comes in contact with stored biosolids may contribute nutrients, pathogens or pollutants. Whether this water accumulates on or near the biosolids, runs off or leaches through the soil, it has the potential to transport contaminants to water resources. Practices to address this issue include (see also Chapter 5 for details and Appendix C):

- Proper shaping of field stockpiles to shed water and avoid puddles of water, or infiltration of water through a stockpile and subsequent loss through runoff or leaching.

- Construction of enclosed storage facilities or tanks.

- Construct lagoons/pads with impervious earthen, concrete or geotextile liners.

- Removal of accumulated water to sites where liquid may be applied.

- Providing buffers between storage areas and waterways.

Preventing Leaching

For permanent, long term storage facilities, an impermeable liner (i.e., earthen, geotextile or concrete) is recommended to ensure against leaching. For all constructed storage facilities, site soils and water table investigations are essential to ensure stable foundations. Soil settling and shifting can result in leakage through cracks. High water tables may float concrete pads or rupture the watertight seals of lagoons.

For short-term field storage, liners are generally unnecessary. Proper shaping of stockpiles encourages shedding of precipitation to prevent infiltration of water and subsequent leaching. Stockpiles should not be located on soils in environmentally sensitive areas with extremely high hydraulic conductivities with excessive infiltration rates, areas with very shallow seasonal high water tables or depths to bedrock, or areas adjacent to or on limestone features such as sinkholes or rock outcrops.

Managing Accumulated Precipitation (See also Chapter 5)

Accumulated water (i.e., precipitation) forms a separate layer on top of liquid or semisolid biosolids or collects in puddles after contact with the material. Overflow or runoff of this water to surface or ground water resources can be prevented or minimized by the following:

- For open storage facilities:
  - use sumps or gravity flow to direct accumulated water to on-site filter strips or treatment ponds,
  - mix accumulated water with biosolids for removal to land application site,
  - decant and transport water accumulations off-site to treatment facilities, or
  - apply to the land through irrigation systems (taking care not to exceed hydraulic loading rates to prevent ponding or runoff).

- For constructed facilities
  - roof to keep precipitation off the material
  - pads should have adequate slope to prevent ponding and appropriate flow management.

Prevent Movement of Biosolids

Direct deposition of biosolids in waterways has the greatest potential for significantly impacting water quality through additions of nutrients, organic matter, pathogens or pollutants. Management practices to prevent this occurrence include:

- Adequate buffers between storage area and water resources.

- Proper storage methods for the physical consistency of the biosolids.

- Proper design and maintenance of constructed storage facilities.

- A spill response and control plan supported by staff training and the availability of the necessary supplies and equipment.

Design and Management Approaches for Water Quality Protection

Proper materials management is an essential measure in water quality protection for all biosolids storage facilities and field stockpiling sites. Well-designed storage operations optimize water quality protection measures by including:

1. structural elements to minimize the potential for accidental spills,

2. operational procedures to reduce potential accidents, and

3. contingency plans to promptly mitigate spills if they do occur (see Chapter 5 for details).

Preventative Measures for Field Stockpiles

- Proper site selection including buffer distances and slopes.

- Proper vehicle and equipment safety features (e.g., waterproof seals on trailer tailgates), maintenance and operator training.

- Adequate staff training and proper operation of site to prevent accidental spills or losses of material to water resources (e.g., truck roll-overs, excess residuals left in loading areas).

- Written spill clean-up and contingency plans and advanced preparation (e.g., equipping storage sites and vehicles with appropriate clean-up tools, and staff drills to ensure rapid and effective response to spills.

Figure 3-1. Staging of biosolids for immediate incorporation into the soil (Maryland.)

Preventative Measures for Constructed Facilities

- Soil strength and suitability assessments prior to construction to avoid uneven settling and other problems that lead to cracks or leaks.

- Adequate design volumes, including space for precipitation accumulations.

- Use of good engineering construction practices to prevent structural failures and malfunctions (e.g., impermeable liners or backflow regulators on gravity systems, paving and curbing of off-loading pads for permanent facilities).

- Proper vehicle and equipment safety features (e.g., waterproof seals on trailer tailgates), maintenance and operator training.

- Adequate staff training and proper operation of site to prevent accidental spills or losses of material to water resources (e.g., truck roll-overs, overtopping of freeboard).

- Written spill clean-up and contingency plans and advanced preparation (e.g., equip sites and vehicles with clean-up tools; conduct staff drills to prepare for effective spill response).

-

Managers of stored biosolids need to assess the nature of their biosolids, the operational requirements and limitations of their land application program, and the storage option most suitable for their operation to select the best combination of design and management practices for their specific situation. To assist in this effort, specific design and management practices for various types of storage options are provided in Chapter 5.

References

CAST. 1996. Integrated Animal Waste Management. Council for Agricultural Science and Technology, Task Force Report No. 128, Ames, IA.

Chaney, R.L. and J.A. Ryan. 1993. Heavy Metals and Toxic organic Pollutants in MSW-Composts: Research Results on Phytoavailability, Bioavailability, Fate, etc., pp. 451-506. *In* Hoitink, H.A.J. and H.M. Keener (eds.), Science and Engineering of Composting. Renaissance Publications, Worthington, Ohio. 728 p.

Gerba, C. P. 1983. Pathogens. *In* A.L. Page, T.L. Gleason, III, J.S., Jr., I.K. Iskandar, and L.E Sommers (eds.) Proceedings: Workshop on Utilization of Municipal Wastewater and Sludge on Land. Univ. of Calif., Riverside, CA.

Hue, N.V. 1995. Sewage Sludge I: Amendments and Environmental Quality, pp. 199-247. *In* J.E. Rechcigl (ed.), Soil Amendments and Environmental Quality. Lewis Publishers, Boca Raton, FL.

Kloepper-Sarns, P., F. Torfs, T. Feijtel, and J. Gooch. 1996. Effects Assessments for Surfactants in Sludge-amended Soils: A Literature Review and Perspectives for Terrestrial Risk Assessment. The Science of the Total Environment 185:171-185.

National Research Council. 1996. Use of Reclaimed Water and Sludge in Food Crop Production. National Academy Press. Washington, D C. 178 pp.

Sharpley, M.A., J.J. Meisinger, A. Breeuwsma, J.T. Simms, T.C. Daniel, and J. S. Schepers. 1998. Impacts of Animal Manure Management on Ground and Surface Water Quality. *In* J.L. Hatfield and B.A. Stewart (eds.). Animal Waste Utilization: Effective Use of Manure as a Soil Resource. Ann Arbor Press, Chelsea, MI. 320 pp.

State of Maryland. 1994. 1994 Maryland Standards and Specifications for Soil Erosion and Sediment Control. MD Dept. Environ. Water Management Admin., Soil Conservation Service and MD State Soil Conservation Committee. Baltimore. MD.

Chapter 4

Pathogens

Introduction

Untreated wastewater contains pathogens, such as viruses, bacteria, and animal and human parasites (protozoa and helminths) which may cause various human diseases and illnesses. Oftentimes these pathogens are or become attached to the separated wastewater solids. It is precisely because of the potential presence of pathogens in untreated wastewater that treatment processes are used to clean wastewater prior to discharge to streams. This is also the reason that wastewater residuals must be subjected to additional pathogen reduction treatment prior to land application of the biosolids.

These treatment processes in the U.S. are carefully regulated and monitored to ensure a consistent level of treatment and pathogen destruction. The combination of treatment and appropriate biosolids management at land application sites has proven to be effective in preventing the transmission of pathogens that can cause disease. Incidents of infectious disease, through either direct exposure or food and/or water pathways, have not been documented from land application of biosolids in the U.S. since this combination of regulated practices has been implemented.

The potential exposure to pathogens during proper biosolids storage is no greater than that associated with direct land application. This chapter describes prudent management practices recommended to safely store biosolids in a manner that limits the potential for transmission of disease agents. Information in this chapter relates to all three Critical Control Points, and especially to Critical Control Point 3.

Biosolids Products Characteristics

Biosolids destined for beneficial use in land application must meet pathogen reduction criteria for either Class A or Class B according to Part 503 rules. Only biosolids intended for and that meet Part 503 criteria for safe land application should be placed in a field stockpile or a constructed storage facility. The two classes of biosolids have different characteristics that influence storage management considerations. Documentation of Class A or B treatment may be achieved either through testing of the final product for specific pathogens or indicator organisms and /or by use of approved treatment processes. Appendix C provides a list of approved Class A and Class B processes.

Class A

Class A biosolids typically are treated by a **"Processes to Further Reduce Pathogens" (PFRP)** such as composting, pasteurization, drying or heat treatment, advanced alkaline treatment, or by testing and meeting the pathogen density limits in Part 503. Class A pathogen reduction reduces the level of pathogenic organisms in the biosolids to a level that does not pose a risk of infectious disease transmission through casual contact or ingestion.

EQ

Class A biosolids which also meet one of Part 503 VAR options 1-8 and meet the metals limits (Part 503 Table 3) are designated as **"Exceptional Quality (EQ)"**. These products are exempted from the Part 503 General Requirements, Management Practices and Site Restrictions, and may be generally marketed and distributed.

Class B

Class B biosolids typically are treated using a **"Process to Significantly Reduce Pathogens" (PSRP)** such as aerobic digestion, anaerobic digestion, air drying, and lime stabilization. As an alternative, producers may document compliance by analyzing the material for fecal coliform levels. When Class B requirements are met, the level of pathogenic organisms is *significantly* reduced, but pathogens are still present. In this case, other precautionary measures required by the Part 503 rule , i.e., site and crop harvesting restrictions, are implemented to protection of public health.

In addition to the pathogen reduction requirement, biosolids must also be treated to reduce their attractiveness to vectors such as rodents, flies, mosquitoes, etc. capable of transmitting pathogens. Part 503.33 of the federal rule specifies analytical standards and treatment processes to achieve Vector Attraction Reduction (VAR) requirements. These include volatile solids reduction, elevation of pH, soil incorporation etc. (see Appendix C).

Biosolids Storage Considerations

Pathogens in Stored Class A Biosolids

In general, storage of Class A biosolids present few pathogen concerns due to the level of pathogen reduction achieved by the treatment processes. The potential for exposure to viruses or parasites (helminth ova) in a Class A product is insignificant as a result of treatment and because these organisms cannot grow outside a suitable host organism. This potential does not increase during storage. Treatment also reduces bacterial pathogens to safe levels. However, bacteria depend on readily available sources of nutrients, adequate water, and favorable environmental conditions, and can grow without a host organism. In specific and very limited situations, the necessary combinations of these factors have been found to occur in stored Class A biosolids. Three examples of these circumstances are:

1. If Class A biosolids compost that is no longer self-heating is blended with green or unstabilized organic materials, such as fresh yard trimmings, fresh hay, or *green* woodchips, the bacterial population can grow rapidly. This is because these fresh materials contain readily available carbon that bacteria need and the compost lacks. If these types of mixtures are managed as self-heating compost piles, i.e., time/temperature conditions adequate to destroy bacterial pathogens are achieved, then the final products will also contain undetectable levels of pathogens as do Class A biosolids. At such low concentrations, disease will not be transmitted even with direct contact with biosolids. If Class A biosolids are mixed with products that contain unavailable carbon sources, such as cellulose and lignin in paper and wood processing residuals, pathogen concentrations will remain undetectable because these nutrients cannot be used by pathogens.

2. If a Class A product is inadequately composted, or its nutrients are not well stabilized bacterial pathogen growth will not occur as long as the material is kept very dry, i.e., total solids content of 80 percent or greater. However, if such dry materials take on moisture during storage, and nutrients, pH, temperature, and other environmental conditions are favorable, pathogen and microbial regrowth could occur. Thus, preparers should be aware that if they conduct various types of blending or permit water content to increase in heat-dried Class A products, the potential for temporary increases in bacterial growth exists

   It is important to recognize that growth during storage is usually a temporary condition in which bacterial populations increase in response to the sudden availability of a food source, but decline to previous low levels once it is consumed. The growth and presence of non-pathogenic microorganisms in biosolids act to counterbalance the stimulating effect of nutrients on bacterial growth through the natural competition for nutrients.

   If pathogen regrowth occurs, the material should be held in storage until populations decline to acceptable levels or it should be re-treated to meet standard pathogen limits. The potential for pathogen growth should be considered in establishing appropriate storage conditions and in blending or augmenting Class A biosolids with other organic materials (see Chapter 7, "Other Organic By-Products").

3. If the pH of Class A alkaline stabilized material drops significantly during extended storage and the color, consistency, or odor of the product has deteriorated, then re-testing for pathogens may be advisable. Significant

decreases in pH have, on occasion, been associated with increases in the level of fecal coliform above the 1000 MPN per gram regulatory limit.

Pathogens in Stored Class B Biosolids

The probable presence of pathogenic organisms is assumed for biosolids treated to Class B pathogen reduction standards. Likewise, Class B biosolids blended with any other organic materials, e.g., leaves, sawdust, woodchips etc., for whatever reason, is not considered to alter the pathogen status. For this reason, storage practices should provide a level of protection equivalent to Class B site restrictions for use to minimize human, animal, or environmental exposure to disease-causing organisms either through direct contact or via the food chain.

PART 503 PATHOGEN DENSITY LIMITS

Biosolids Pathogen Standards can be satisfied by determining the geometric mean of seven samples of biosolids after treatment for the following:

Pathogen or Indicator	Standard density limits (dry wt)	
	Class A	
• Salmonella	< 3 MPN / 4 g Total Solids	or
• Fecal Coliforms	< 1000 MPN/g	and
• Enteric Viruses	< 1 PFU / 4 g Total Solids	and
• Viable Helminth Ova	< 1 / 4 g Total Solids	
	Class B	
Fecal Coliform Density	<2,000,000 MPN/ g Total Solids (dry wt. basis)	

Accumulated Water

Ponded water that has contacted stored biosolids may contain nutrients and have a moderate enough pH to provide a favorable medium for growth of bacteria, including pathogens. This may occur even when the bulk of the stored product is dry. In addition, according to the preliminary risk assessments for land application of biosolids, the highest risk pathways for viruses, bacteria and parasites involve direct human contact with biosolids or with surface waters that have been contaminated by runoff and sediment, particularly immediately after a rainfall. Therefore, management of stormwater to minimize contact with biosolids and properly dealing with any water that accumulates in contact with stored biosolids is essential.

When is Retesting Required?

Class A and EQ

For EQ biosolids the Part 503 requirements to test stored materials prior to use depends on who has control of the stored material. If the material remains in the control of the original preparer (directly or indirectly through a contracted processor or applier), the material must be retested prior to final use. If a preparer gives or sells EQ biosolids to a second party, for instance a landscaper, who then stores the material before land application, testing for pathogens is not required under Part 503.

The two examples above are often referred to as the "quirk" of the EQ concept. In one case, the EQ biosolids is still subject to the Part 503 requirements when something happens to it because it is still *under the control of the preparer*. In the other case, the same EQ biosolids is not subject to the Part 503 requirements when something happens to it because it is *no longer under the control of the preparer*. Loss of control by the preparer is the critical difference conceptually. However, even second party receivers of EQ materials should be aware that pathogen testing is recommended when bulk blending operations of biosolids with materials that contain available nutrients occur.

Class B Material

For Class B biosolids, any mixture of a Class B biosolids and a non-hazardous material is considered as a product derived from biosolids, and hence, by definition, biosolids. Thus, if either a preparer or a land applier blends ground green waste with Class B biosolids, and then plans to till that mixture into the soil the mixture would still need to meet the Part 503 Class B standard and site restrictions (i.e., pollutants, pathogen, and vector attraction reduction requirements). The party who mixes the biosolids with another material is the preparer, as defined in Part 503.

Land appliers who are considering or are already blending biosolids with other materials prior to ultimate disposition of the product need to be aware of Part 503 requirements for biosolids derived products. This means that if the blends with Class B biosolids are stored, when they are removed from storage for land application, they must still use site restrictions.

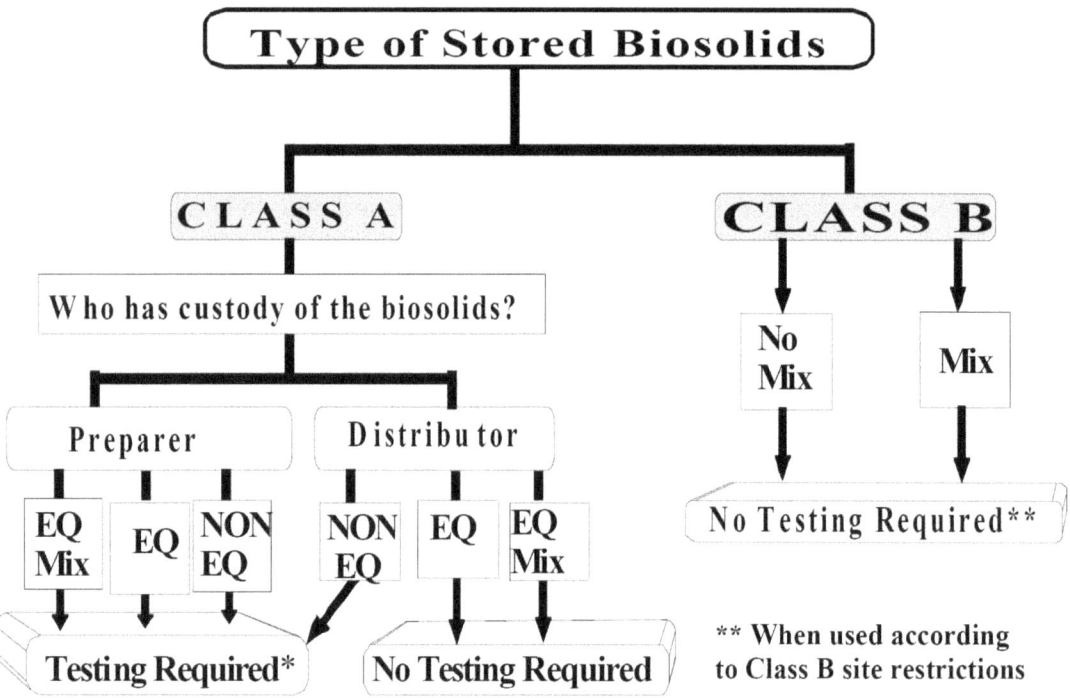

* Before custody of the biosolids is transferred to the distributor, OR
when something other than EQ biosolids is mixed with NON-EQ
biosolids after the preparer has released control of it.

If anything is mixed with NON-EQ biosolids, the mixture is subject to
the land application general requirements and management practices
when it is land-applied.

Fig. 4-1. Decision tree diagram showing the interrelationship between biosolids
pathogen reduction status (Class A, B, and EQ), current custodian, and mixing
with non biosolids material relative to testing and retesting requirements.

Storage Site Management

Three conditions are necessary to produce infectious disease:

• The disease agent must be present in sufficient concentrations to be
infectious
• Susceptible individuals must come in contact with the agent in a manner that
causes infection
• The agent must be able to overcome the physical and immunological
barriers of the individual.

Proper management practices break the chain of transmission either by
keeping susceptible individuals or animals from direct contact with stored

materials and/or by preventing the movement of any residual pathogens or parasites in stored materials into the environment in a way that would be harmful. Biosolids regulations are designed to address the first two of three conditions that produce infections disease.

- Biosolids which meet rigorous Class A pathogen reduction standards do not have detectable levels of pathogens and are exempt from site restrictions.
- For Class B biosolids, the risk of transmission of infectious disease agents is reduced to acceptable levels by a combination of treatment to reduce pathogen levels and management practices to minimize the potential for exposure of susceptible individuals to pathogens or parasites.

Management Options to Restrict Potential Movement of Pathogens

- Use of appropriate buffers or filter strips to control runoff from field stockpiles.

- Diverting stormwater runoff away from the stored biosolids.

- Practices such as stormwater containment ponds or collection and irrigation systems for uncovered constructed storage pads or lagoons.

- Enclosure of long term storage of biosolids in facilities with roofed structures to prevent contact with precipitation or runoff where feasible.

- Restriction of public access to field storage sites. Constructed facilities may warrant fencing, but fencing of field storage stockpiles is needed only if storage will occur in areas that are accessible to livestock.

- Any runoff which has been in contact with the biosolids should be kept isolated from any adjacent fruit or vegetable crops that would be harvested, sold in the fresh market, and potentially consumed raw.

Chapter 5 includes detailed discussion of management practices that minimize pathogen transport or exposure risks for a variety of biosolids storage options.

Worker Safety

Worker safety is always a primary consideration and basic hygiene training similar to that of workers at a wastewater treatment plant should be provided to biosolids haulers and storage site staff. The use of good personal hygiene and work habits form the basis of a worker protection program for those handling biosolids. Some specific recommendations include:

1. Wash hands thoroughly with soap and water after contact with biosolids.

2. Avoid touching face, mouth, eyes, nose, genitalia, or open sores and cuts.

3. Wash your hands <u>before</u> you eat, drink, smoke, or use the restroom.

4. Eat in designated areas away from biosolids handling activities.

5. Do not smoke or chew tobacco or gum while working with biosolids.

6. Use gloves to protect against creation of skin abrasions and/or contact between abrasions and biosolids, or surfaces exposed to biosolids, when they occur unexpectedly.

7. Remove excess biosolids from shoes prior to entering vehicle.

8. Keep wounds covered with clean, dry bandages.

9. Flush eyes thoroughly, but gently, if biosolids contact eyes.

10. Change into clean work clothing on a daily basis and, if possible, before going home; reserve work boots for use at storage sites or during biosolids transport.

The Centers for Disease control recommends that immunizations for diphtheria and tetanus be current for the general public, including all wastewater workers. Boosters are recommended every ten years. The tetanus booster should be repeated in the case of a wound that becomes dirty, if the previous booster is more than five years old. Consult a doctor regarding direct exposure through an open wound, eyes, nose, or mouth. It should be noted that a Hepatitis A vaccine has recently been developed and is available to the general public. Consequently, it is recommended that those working with biosolids receive this vaccination as an additional protection.

References

Code of Federal Regulations, 1993. Standards for the Use and Disposal of Sewage Sludge. Title 40, Volume 3, Parts 425 to 699, Federal Register February 19, 1993 (58 FR 9248), US Government Printing Office, Washington, DC [40CFR503.3].

EPA, 1992. Environmental regulations and technology - control of pathogens and vector attraction in sewage sludge, EPA Pub. No. 625/R-92/013, Center for Environmental Research Information, Cincinnati, OH 45268.

EPA, 1992b. Preliminary Risk Assessment for Viruses in Municipal Sewage Sludge Applied to Land. EPA Pub. No. 600/R-92/064, EROC/CSMEE, Columbus, OH.

EPA, 1991a. Preliminary Risk Assessment for Bacteria in Municipal Sewage Sludge Applied to Land. EPA Pub. No. 600/6-91/006, EROC/CSMEE, Columbus, OH.

EPA, 1991b. Preliminary Risk Assessment for Parasites in Municipal Sewage Sludge Applied to Land. EPA Pub. No. 600/6-91/001, EROC/CSMEE, Columbus, OH.

EPA, 1989. Environmental regulations and technology - control of pathogens in municipal wastewater sludge, EPA Pub. No. 625/10-89/006, Center for Environmental Research Information, Cincinnati, OH 45268.

EPA, 1989. Technical support document for pathogen reduction in sewage sludge. Publication no. PB 89-136618. National Technical Information Service, Springfield, Virginia.

EPA, 1985. Health effects of land application of municipal sludge. EPA Pub. No. 600/1-85/015. EPA Health Effects Research Laboratory, Research Triangle Park, North Carolina.

EPA, 1979. Technology Transfer Process Design Manual - Sludge Treatment and Disposal, EPA 625/1-79-011, Center for Environmental Research Information, Cincinnati, Ohio.

Farrell, J.B., V. Bhide, and J.E. Smith, Jr., 1996. Development of EPA's new methods to quantify vector attraction of wastewater sludges. Water Environ. Res. 68, No. 3, 286-294.

Feachem, R.G., D.J. Bradley, H. Garelick, and D.D. Mara. 1983. Sanitation and disease: health aspects of excreta and wastewater management. Wold Bank Studies in Water Supply and Sanitation 3. John Wiley & Sons, New York.

Smith, J. E., Jr., and J. B. Farrell. 1994. Vector Attraction Reduction Issues Associated with the Part 503 Regulations and Supplemental Guidance. in Proceedings of the Water Environment Federation's Conference, "International management of water and wastewater solids for the 21st century: A global perspective", June 19-22, 1994, Washington, D.C., pp 1311-1330.

Strauch, D. 1991. Survival of pathogenic microorganisms and parasites in excreta, manure and sewage sludge. Rev. Sci. Tech. Off. Int. Epizoot. 10(3):813-846.

Yanko, W.A., A.S. Walker, J.L. Jackson, L.L. Libao, and A. L. Gracia. 1995. Enumerating Salmonella in biosolids for compliance with pathogen regulations. Water Environ. Res. 67(3): 364-370.

Chapter 5

Recommended Management Practices

Introduction

This chapter deals with the various issues of <u>Critical Control Point 2</u>: *The Transportation Process* and <u>Critical Control Point 3</u>: *The Field Storage Site*. Design guidance and management recommendations are provided for storage of biosolids that meet state and federal standards and are suitable for use in land application programs. The operative concept for these recommendations is that site design and management requirements increase as the length of storage or volume of stored biosolids increases. These recommendations are based on practical field experience and are designed to protect water quality, minimize pathogen exposure risks, and reduce the potential for unacceptable off-site odors.

The five sections in this chapter are:

I.   Site Selection Considerations: Applicable to All Storage
II. Field Storage: Stockpiles
III. Field Storage: Constructed Facilities
IV. Odor Prevention and Mitigation
V.   Spill Prevention and Response

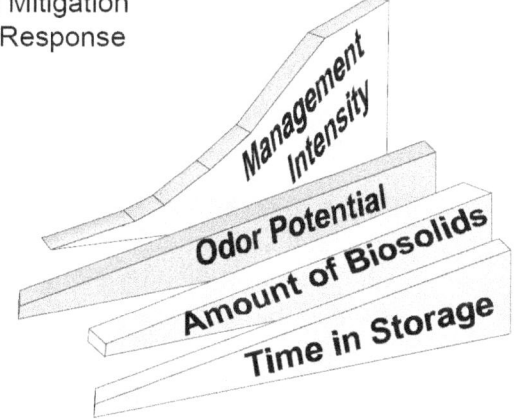

I. Site Selection Considerations: Applicable to all storage

> *SITE SELECTION FACTORS*
>
> **CLIMATE**
>
> **TOPOGRAPHY**
>
> **SOIL/GEOLOGY**
>
> **BUFFER ZONES**
>
> **ODOR PREVENTION/AESTHETICS**
>
> **ACCESSIBILITY AND HAULING DISTANCE**
>
> **PROPERTY ISSUES**

Climate

The climate of the area should be assessed to determine the likelihood of precipitation events over the planned storage period, the expected temperatures, wind speed and prevailing seasonal directions relevant to sensitive odor receptors. For constructed facilities, the anticipated length of inclement weather conditions and rainfall may influence the size of the facility. For instance, in many areas of the United States, land application of biosolids is severely limited from the months of November through March.

Topography

Field stockpiles and storage facilities should not be located in areas that are regularly inundated, in drainage ways or in wetlands. They should be placed on fairly level land. Stockpiles should be situated near the top of slopes to minimize exposure to up-slope runoff. Constructed storage facilities may require storm water controls if subjected to up-slope runoff. U.S. Geologic Survey (USGS) topographic maps are an excellent tool for screening of suitable locations. Biosolids should be stored in areas with adequate buffers.

Soils and Geology

Sites selected for field storage should not be located on excessively moist or wetland soils where very low infiltration rates regularly lead to standing water or excessive runoff after storm events. Stockpiles also should not be located on soils with extremely high hydraulic conductivities (such as gravels) that have excessive infiltration rates. Regulatory requirements and water quality protection standards regarding depth to seasonal high water table and to bedrock should also be considered. Stockpiles do not belong on or adjacent to karst features such as sinkholes or rock outcroppings.

For constructed storage facilities, the soil must provide a suitable foundation. The movement, settling and shifting of the underlying soil could result in leakage through cracks or even the total failure of the storage structure. High water tables may pose the risk of rupturing the water-tight seals of a lagoon (particularly with clay lined systems) causing groundwater infiltration into the storage facility or conversely leakage of the biosolids to the surrounding groundwater. High permanent or seasonal water tables may also exert enough flotation force on concrete or steel structures to lift them from their foundations. The soil at the site should be evaluated in regard to its suitability and strength for use in embankments, berms and backfill. It may be necessary to truck in suitable soils, which will significantly increase the cost of the storage facility.

Buffers

Adequate buffers are necessary to protect water resources and to prevent nuisances to adjacent properties. The storage site should comply with any federal (10 meters by the 503 rule), state, or local regulations regarding minimum buffer distances to waterways, homes, wells, property lines, roads, etc. They also prevent surface runoff from reaching streams by providing room for infiltration in crop areas and vegetative buffers or crop residue.

Odor Minimization and Aesthetics

Reducing the visibility of the storage site to the general public and maximizing the distance between the site and residential areas help minimize nuisance complaints. The length of time biosolids are stored should be minimized when sites are adjacent to residential areas. Storage during the summer months poses a greater potential for development of unacceptable odors and requires a higher level of management.

Accessibility and Hauling Distances

Potential sites should be evaluated based on economical hauling radii from the generating facility and the accessibility of the site during periods of inclement weather. Weight restriction and other roadway limits should be observed both on-site and along the haul route from the treatment facility. Consideration should also be given to traffic impacts on communities along the haul route and the least disruptive route selected.

Biosolids must be transported to the storage site in vehicles that are appropriate for the type of materials being transported, e.g., for dewatered or dried biosolids, trucks must be covered and have rubber sealed rear gates. Drivers should be briefed on haul routes and provided with a copy of a written spill response plan that describes emergency response and clean-up methods in the event of a spill, accident etc. It is advisable to keep one in each haul vehicle and at project offices. Investigate and comply with any local road use requirements or restrictions.

Prior to biosolids deliveries, mark field stockpile areas. Place signs or flags conspicuously enough for truck drivers to determine where to drive and unload biosolids. Make provisions for collection of load tickets to document deliveries.

Permanent storage facilities must have room for trucks to maneuver and pull-offs or staging areas to ensure vehicles do not queue up on the shoulder of public roads while waiting to be loaded or unloaded. In landscape and horticultural uses, where Class A biosolids will be combined with other materials, consider locating storage areas near other amendments to minimize the time required to collect and blend potting and landscaping mixtures.

Property Issues

Before constructed storage facilities are built, local zoning requirements and ordinances must be investigated. In addition, consideration should be given to the relative security and liability associated with leasing versus ownership of the land on which the storage facility will be located. Any leases should extend for several years and preferably over the expected life of the facility. Leases should have provisions that allow and guarantee proper management of the site and compliance with regulatory requirements. Plans should also be made for the eventual closure of the facility such as demolition and restoration of the site or conversion of the facility to other uses. Adequate insurance of the property, facility and equipment as well as environmental liability coverage is necessary. This coverage must be coordinated with any applicable state or local bonding requirements.

II. Field Storage: Stockpiles

Field Storage Considerations for Stockpiles

DESIGN CONSIDERATIONS

SITE SELECTION & WATER MANGAGEMENT

OPERATIONAL PRACTICES

HOUSEKEEPING

SECURITY

SITE RESTORATION

A *Critical Control Point 3: Field Storage (Stockpile) Checklist* (page 48) summarizes material discussed in this section.

Design Considerations

Field stockpiling is used for short-term storage of dewatered cake, dried, or composted Class B or Class A biosolids at the land application site. Use biosolids that stay consolidated and non-flowing -- It is advisable to test the biosolids' ability to stay consolidated before field stockpiling operations are initiated; such testing can be conducted at the treatment plant. This should be rechecked if a treatment plant changes polymers or dewatering methods. This test is suggested because some polymers used in dewatering may break down after a couple days. If this occurs, bound water in the biosolids is released and the stockpiled biosolids may lose solidity and slump or flow.

If biosolids do not have the proper consistency, they may be blended with thicker biosolids from the treatment facility. If Class A is mixed with Class B, the material must be handled as a Class B biosolids unless the mixture is retested and/or retreated to meet Class A standards.

Alternatively, it may be feasible in some situations to stockpile biosolids on a layer of sawdust or other absorbent material. Such practice is not considered to change the quality of the biosolids, and hence does not require a federal "Treatment Works Treating Domestic Sewage (TWTDS)" application and additional testing for Part 503 compliance.

Site Selection and Water Management

Field stockpiles should be placed in the best physical location possible in or adjacent to the field(s) that will receive the biosolids. Stockpiles should be placed according to the general siting recommendations listed earlier in this chapter and conform to all state requirements. For sites with significant slope, provisions need to be made to manage up-and downslope water. Avoid forming windrows across slopes to reduce the potential for piles to become anaerobic at the base where overland flow accumulates. To the extent possible, shape piles to shed water. Clearly mark access routes and stockpile areas at field sites.

Upslope
The longer the storage period, the greater the potential precipitation, and hence, greater levels of runoff control are needed. Runoff from any up-slope areas should be diverted by using straw bales, silt fence, by discing soils up-slope of the stockpile along the contour line, or by constructing a berm with soil from the site. In some cases, inert, low nitrogen, residuals such as agricultural lime, pulp/paper sludge, or wood ash have been used successfully as berm materials. For schematic diagrams of several types of berm construction see Appendix C.

Downslope

Ensure that measures are placed down-slope of the stockpile to manage runoff. These measures could include buffers or filter strips consisting of established vegetation or crop residues, tillage across the slope to increase soil roughness, silt fencing, straw bales, or berms (see Appendix D for schematic details on construction). The extent of these measures should be based on the length of time the material is expected to remain stockpiled and the likelihood of significant runoff events occurring during this period. The amount of biosolids stored at a field stockpile site should be limited to that which can be used on the adjacent fields.

Covering

Stockpiled biosolids form an air-dried crust that sheds precipitation and prevents significant percolation of water through the pile. Nonetheless, some states require stockpiles to be covered. However, field experience has shown that tarps are not practical, except for very small stockpiles.

Biosolids stockpiles usually occupy a significant area; large tarps needed to cover them are expensive, difficult to anchor and handle. Spreading the tarp often requires workers to physically wade in biosolids. Furthermore, placing and removing tarps may lead to significant drag-out of biosolids and the soiled tarps themselves are a disposal problem. Shredded bark, compost, straw mulch, ash, or topical lime application at times have been used as covers for biosolids stockpiles (primarily to minimize odors as necessary).

For dried (at least 50 percent solids) or composted biosolids, tarps, wind barriers or periodic wetting may be necessary to minimize blowing of dust, particularly in arid, windy, climates when stockpiles are in close proximity to sensitive downwind areas, e.g., residential areas. There have been some instances of tarps catching fire when used on compost materials. Hence, monitoring for hot spots as described below in 'inspections' is a useful preventive strategy.

On a practical basis, several methods of effectively minimizing potential water quality impacts include proper shaping of stockpiles, whenever possible, to shed water, up-slope runoff diversions, and down-slope filter strips or other practices. For biosolids-derived materials slated for use in highway projects, consider storing the material on paved surfaces below overpasses to shelter from precipitation.

Operational Practices

Inspection

Stockpiles should be inspected regularly and after severe precipitation events to ensure that runoff controls are in good working order; note any slumping, erosion, or movement of the biosolids; ensure there is no ponding or excessive odor at the site. It is recommended that an inspection report be completed, documenting the time, date, person conducting the inspection, and any items requiring maintenance or repair.

Compost Inspection
Incompletely composted materials have the potential to self-heat because microbial growth can still occur on the remaining nutrients. Thus, it is important that only stable compost product be placed in storage. The compost reheat test is an easy, on-site test that a compost producer can use to determine if this level of compost stability has been achieved prior to storage. Alternatively, there are oxygen uptake and carbon dioxide test prosedures that can be used at the production facility. Temperature of stockpiles can be monitored conveniently and rapidly with hand-held, 'point and shoot' infrared temperature devices approved by the National Fire Protection Association to ensure the material does not become a potential fire hazard. Steel temperature probes inserted into various places in a pile for approximately 10 minutes can also be used, as can thermistor probes that are buried in piles and relay temperature data to a remote, electronic data-acquisition system.

Heat Dried Product Inspection
Heat dried products that are rewetted or have not been sufficiently dried and cooled (<95% solids, >85°F) also can self-heat. In the presence of enough available water, microbes will utilize the nutrients in the biosolids and generate heat that cannot dissipate because of the mass of the stockpile. Therefore, piles should be monitored if rewetting occurs so that a fire hazard does not develop. A noticeable increase in odor is a reliable indicator of microbial activity and the potential development of hot spots. Temperature monitoring devices as used for composting can also be used with stored, heat-dried biosolids. If hot spots are found, the stockpile should be broken apart to vent the heat and dry, then restacked or the material should be land applied.

In arid regions or during droughts, prudent management practices for potentially combustible material might also include:

- A fire break of 30 ft. around stored materials by removing combustible vegetation
- Foam-type fire suppressant or emergency water source (tank), possibly including detergent to enhance the surface contact effectiveness of the water.

Housekeeping

During stockpile creation or removal, employees must ensure that, at the end of each work session, runoff controls are in place and stockpiles are properly shaped whenever possible to prevent ponding of water on top of the biosolids. They must also ensure that equipment is clean and the area is secured. For a list of practices useful in preventing the tracking of mud and biosolids onto public roads see p. 58.

Security

Locate field storage piles in remote areas of sites, when possible, to limit access. Install appropriate fencing around stockpiles located on fields where livestock will be grazed during the storage period to prevent their access.

Site Restoration

For most soils, stockpile areas exhibit soil compaction (especially when wet) due to heavy equipment operation. Evaluate soil sensitivity to compaction when selecting the loading and storage areas. Storage areas may also exhibit high levels of nutrients, salts, and pH (for limed biosolids), that may potentially inhibit seed germination and crop growth. For these reasons, the following measures are often needed after biosolids are removed from a stockpile and land applied:

- Remove and spread the residual biosolids in the stockpile area. This can be accomplished using a loader bucket to closely skim biosolids from the ground surface and, if necessary, dragging the area with the back of the loader bucket. In some cases, where equipment has churned biosolids into the soil, it may be advisable to scrape a thin layer of soil with the biosolids. Where biosolids are stockpiled on hay or pasture, it may be necessary to use a chain drag to breakup and spread out biosolids left in the loading area.

- When cropping practices allow, the soil in the stockpile area should be tilled with a disc, chisel plowed, subsoil tilled etc., to breakup compaction. The site should then be seeded or cropped to take up nutrients. If there are several suitable locations at a site, stockpiles should be rotated from year to year rather than repeatedly placed in the same location. If a single area of the site will be used repeatedly, this area will need a higher level of management.

	Critical Control Point 3: Field Storage (Stockpile) Checklist	
	(Involving dewatered cake, dried, or composted Class A or Class B Biosolids)	
	Management	✔
1	Prepare and maintain a Field Management Plan	
2	Train employees to properly operate the site according to plan; conduct spill drills	
3	Critical Control Point 1: Work with WWTP to maximize biosolids stability, consistency, and quality; direct batches to appropriate sites.	
4	Critical Control Point 2: Transportation; Clearly mark site access routes and stockpile areas; conduct spill drills	
5	Maintain accurate and well organized records	
6	Designate a competent public relations person; maintain communication with stakeholders; notify agencies of reportable incidents; explain actions taken to respond to citizens concerns or complaints	
	Operations	✔
1	Use biosolids that stay consolidated and non-flowing; shape stockpiles whenever possible to shed water	
2	Minimize ponding and storage time to the extent feasible during hot, humid weather; manage accumulated water appropriately	
3	Inspect and maintain up-slope water diversions	
4	Inspect buffer zones to ensure run-off is not moving out of bounds	
5	Restrict public access and use temporary fencing to exclude livestock, where applicable; install signs; secure site appropriately	
6	Clean all vehicles and equipment before they exit onto public roads	
7	Train employees to use of appropriate sanitation practices; inspect for use	
8	Inspect for odors and conditions conducive to odors; apply chemicals or surface covering material to suppress odors if needed; consider the meteorological conditions and the potential for off-site odors when scheduling opening the storage pile and spreading of biosolids	

Figure 5-1. Daily biosolids deliveries are temporarily stored in a steel box fabricated from two intermodal freight containers (Snoqualmie Tree Farm, WA). The box breaks down and stacks together lengthwise for relatively easy relocation to the next unloading site. Biosolids are loaded from the containers into the Aero-Spread applicator by a clam bucket.

Figure 5-2. Temporary stockpiles of biosolids in Maine covered with lime mud (high pH) that acts as an odor control measure until material is incorporated. (Courtesy of Mark King, Maine Dept. Environmental Protection)

III. Field Storage: Constructed Facilities

Introduction

A checklist of the items discussed in this section appears on page 59.

Longer-term storage is often conducted at constructed facilities where additional steps and management practices have been implemented to protect human health and the environment. Constructed facilities include: concrete, asphalt, clay, or compacted earth pads; lagoons; tanks; or other structures that can be used continually to store liquid, semi-solid or solid biosolids. Generally these facilities are made of impervious materials that prevent leaching and have specific design components to manage precipitation and runoff.

Design and management options presented here for short- or long-term storage of biosolids, are based on current technologies and actual experiences. These options are not the only effective ones. New or innovative options may provide equal or better management.

Design Considerations

Field stockpiling is generally limited to the amount of biosolids needed to meet agronomic or reclamation requirements at a field or site. Determining the storage period and suitable capacity for a constructed facility is more variable, and is a critical component of most well managed land application programs. If the capacity is for too short a period, the facility may fill before the biosolids can be used in a sound manner. A design that is based on an overly long storage period may result in an unjustifiable expenditure for unused storage capacity.

Factors to consider in determining the storage period include the daily production at the WWTP, storage alternatives, climate and land use characteristics, equipment and labor requirements, and management flexibility. The larger the capacity for storage, the greater the flexibility in managing biosolids to accommodate weather, equipment, etc.

Constructed facilities should be designed and built in accordance with good engineering principles. Excellent guidance on these types of facilities is available in the Natural Resources Conservation Service (NRCS) design manuals for animal manure storage facilities. State and local regulatory requirements and design criteria provide details. The time vs. amount vs. management intensity relationship applies to these facilities as much as it does to stockpiles. Table 5-1 provides key design considerations for the three types of constructed facilities customarily used to store biosolids products:

- Lagoons for liquid or dewatered biosolids
- Pads or other facilities for dewatered or dry biosolids
- Storage tanks for liquids

Table 5-1:   Key Design Concepts for Constructed Biosolids Storage Facilities

Issue	Liquid/ Thickened 1-12% solids	Dewatered/Dry Biosolids Facilities 12-30% solids/ >50% solids (dry)		Liquid/ Thickened 1-12% solids
	Lagoons	Pads/Basins	Enclosed Buildings	Tanks
Design	Below ground excavation. Impermeable liner of concrete, geotextile, or compacted earth.	Above ground, impermeable liner of concrete, asphalt, or compacted earth	Roofed, open-sided or enclosed. Flooring: concrete, asphalt, or compacted earth	Above or below ground, concrete, metal or prefab. If enclosed - ventilation needed
Capacity	Expected biosolids volume + expected precipitation +freeboard	Expected biosolids volume, unless precipitation is retained; then, biosolids volume + expected precipitation +freeboard	Expected biosolids volume	Enclosed: expected biosolids volume. If open-top - expected biosolids volume + expected precipitation +freeboard
Accumulated Water Management	Pump out and spray irrigate or land apply the liquid, haul to WWTP, or mix with biosolids	Sumps/pumps if facility is a basin for collection of water for spray irrigation, land apply or haul to a WWTP	Roof and gutter system, enclosure, or up-slope diversions	Decant and spray irrigate, land apply or haul to WWTP or mix with biosolids in tank
Runoff Management	Diversions to keep runoff out of lagoon	Diversions to keep runoff out of site, curbs and/or sumps to collect water for removal or down-slope filter strips or treatment ponds	Enclosure or up-slope diversions	Prevent gravity outflows from pipes and fittings. Diversions for open, below ground tanks
Biosolids Consistency	Liquid or dewatered - removal with pumps, cranes or loaders	If no side-walls, material must stack without flowing	Material must stack well enough to remain inside	Liquid or dewatered biosolids. If enclosed, material must be liquid enough to pump.
Safety	Drowning hazard - post warnings, fence, locked gates and rescue equipment on site	Drowning hazard - post warnings, fences, locking gates, and rescue equipment on site	Post 'No Trespassing',\ signs, remote location, lock doors, gates & fences	Posted warning., locking access points, e.g., use hatches, controlled access ladders, and confined space entry procedures to access

Lagoon Storage

Storage lagoons need to be large enough to provide adequate biosolids storage volumes during worst-case weather conditions (long periods of inclement weather when field application is restricted and the lagoon storage cannot be emptied). The design volume must also include space for accumulation of precipitation expected over the storage period plus capacity to hold severe storm events (e.g., a 2-year, 24-hour design storm). Lagoons must also have adequate freeboard (the distance from the maximum water level to the top of the berm).

An impermeable liner (i.e. earthen, geotextile, or concrete) is recommended to ensure against loss of biosolids constituents to groundwater by leaching. This type of design may negate the need for groundwater monitoring wells. Liners should be protected from damage by restricting vehicle access to concrete ramps and vehicle lanes. If vehicles must traverse the liner surface or if dredges will be used to remove biosolids, a layer of sand (approximately one-foot thick) or clay should be spread over the liner. This sand or clay base is protection in itself and provides a marker to indicate when removal operations are approaching the liner.

Fig. 5-5 A lined lagoon *(Courtesy of BioGro Division).*

Dewatered/Dry Biosolids Storage Facilities
Dewatered/dry biosolids storage facilities can be covered or uncovered and are designed to provide up to two years of storage for Class A or B dewatered, air-dried, heat dried, or composted biosolids. These facilities include open-sided or enclosed buildings and open topped bunkers or pads. Storage facilities need to be large enough to provide adequate biosolids storage volumes during worst-case weather conditions (long periods of inclement weather when field application is restricted and the facility cannot be emptied). If the facility is not under roof, the design must provide for stormwater retention apart from the stored biosolids with sufficient volume for precipitation accumulation or provide other management measures that prevent accumulation.

Unroofed facilities for semisolid cake materials (Class B or Class A with less than 50 percent solids) should have a durable hard pad with push walls and stormwater curbs, containment walls, and sumps. An impermeable floor is recommended to help control runoff, protect against loss of biosolids constituents to groundwater by leaching, and to accommodate vehicle traffic. Recommended materials include concrete or asphalt in humid areas; arid areas may also use compacted soils. Class A material with greater than 50 percent solids (compost, alkaline stabilized etc.) may be stored on bare ground or gravel with appropriate runoff controls, such as straw bales, sediment fence, and grassed filter strips. Facilities with roofs or impermeable floors, when accompanied by appropriate stormwater management provisions protect groundwater.

Figure 5-4. Concrete storage bunker with block push walls (Courtesy Mark King, Maine Dept. of Environmental Protection).

Figure 5-5. Permanent covered storage in Southern Maine (Courtesy Mark King, Maine Dept. Environmental Protection).

Storage Tanks
Storage tanks for Class A and Class B liquid biosolids may be temporary or permanent, above- or belowground structures. They are watertight and are generally concrete or steel structures, which may be prefabricated or constructed entirely on-site. Due to their impervious nature, these facilities generally do not warrant groundwater-monitoring wells -- particularly aboveground tanks.

Storage tanks may be open-topped or enclosed. Like lagoons, open-topped storage tanks must include space for expected precipitation accumulations, plus adequate freeboard. The tank volumes need to be large enough to contain daily biosolids produced during worst-case periods of inclement weather, or, back-up options must be part of the planning process.

Ventilation
Enclosed storage tanks should be ventilated through passive vents or mechanical fans. Depending on the type of biosolids, tank design, climatic conditions, and airflow rates, a gas meter and alarm system tied to ventilation fans may be advisable to eliminate buildup of explosive levels of methane that might result from anaerobic biological activity in the tank. Specific requirements for ventilation and electrical systems on or in the immediate vicinity of different types of enclosed storage facilities are specified in the National Electric Code requirements adopted by the National Fire Protection Association. Post "No Smoking" and "Confined Space" signs on all enclosed storage tanks.

Spills
Aboveground tanks have the potential for spills due to gravity flow of biosolids. Two approaches to protect from accidental spills are:

1. The tank may be designed so that valves and piping on the tank do not allow material to flow out by force of gravity (top feeding systems). In these systems, biosolids are lifted in and out of the tank by pumps. This prevents spills in the event that a valve is damaged by equipment or if an operator fails to shut the valve.

2. For gravity discharge systems, backflow prevention and emergency cut-off valves should be installed on all piping and valves located at elevations lower than the highest potential liquid level of the tank.

Berms
An earthen containment berm may be advisable if the facility is located fairly close to a drainage-way, surface waters, or other sensitive feature. The containment berm should be designed to retard the movement of biosolids spilled from a tanker truck, handling equipment, or the tank itself. The containment berm should detain a spill long enough for it to be cleaned up but include a dewatering device that will prevent ponding of rainwater (see Appendix D for diagrams of berms).

More on Water Management

Surface Runoff/Erosion Controls
During Construction - Control of stormwater and runoff during construction of storage facilities is essential and may be regulated by federal, state or local erosion and sediment control and stormwater regulations. Erosion and sediment controls may include installation of up-slope runoff diversions to keep stormwater from crossing the construction area and by installation of silt fence or other structures along the lower perimeter of the disturbed area to trap stormwater and/or sediment. Areas disturbed during construction should be stabilized to prevent erosion by seeding and mulching.

After Construction - Depending on the type of constructed facility, it also may be necessary to install permanent diversions to keep up-slope surface runoff from entering facilities and other down-slope water management structures. Specifications for erosion and sediment control practices are available at local planning offices and Natural Resources Conservation Service (NRCS) offices. (See Appendix C).

Management of Accumulated Water
Accumulated water (i.e., precipitation) that forms a separate layer on top of liquid or semi-solid biosolids, or collects in puddles after contact with biosolids, is the primary cause of odors at storage facilities. There are two design approaches, prevention and mitigation, for dealing with water accumulation at storage facilities constructed for dewatered and dry biosolids:

Prevention - Construct roofed facilities to prevent water or precipitation from contacting biosolids, and provide additional water management as needed.

Mitigation -
1. Construct curbs, gutters, and sumps at unroofed facilities to collect and manage water that has come into contact with the biosolids; treat such water as liquid biosolids; and/or,
2. Establish gravity flow to on-site filter strips or treatment ponds. In arid regions of the U.S., accumulated precipitation may not need to be managed due to evaporation deficits; and/or,
3. Mix accumulated water with the biosolids, or decant it from the storage facility as quickly and regularly as possible -- especially during warm weather. Use an irrigation system or truck spray system for land application or back haul to the treatment facility (this option may be complicated by expensive tip fees or treatment plant acceptance limits on BOD and nitrogen concentrations).
4. Application to land should be based on nutrient loading rates and hydraulic loading limits to prevent ponding or runoff to adjacent land.

Land application of accumulated water should be treated under state and federal regulations as liquid biosolids, if the water has come in contact with biosolids, and all biosolids management practices and site restrictions should apply. State nutrient management plan requirements will specify nutrient testing. In the absence of state requirements, nutrient testing is recommended. When planning to irrigate accumulated water make sure that adequate land will be accessible when it is needed. Also, check state and local regulations regarding land application in the winter.

Effects of Storage: Application Rate Adjustments

The longer biosolids are stored, the more important it is to retest for nutrients. Before removal, biosolids should be sampled and tested for nitrogen, phosphorus, and percent solids. Liquid biosolids increase or decrease in percent solids over time due to precipitation additions or evaporation losses. In addition, settling may occur during storage. Depending on the degree of liquid/solids separation and the amount of recirculation and remixing that can be achieved, the percent solids of the material may vary from the surface to the bottom of the lagoon. Therefore, it is advisable to retest the percent solids of the material as the clean-out proceeds to ensure proper application rates.

Operational Practices for Constructed Facilities

Inspections
Inspections should be regularly scheduled while biosolids are stored in facilities to determine if any maintenance or repairs are necessary. The site should also be checked for odors, proper management of precipitation, housekeeping, and security. Inspections after rainfall events during periods of warm weather are particularly helpful in preventing the development of unacceptable odors. An

inspection report should be completed, documenting the time, date, person conducting the inspection, and any items requiring maintenance, repair, or adjustment.

Visual inspections should include examination of the condition of:

- **Liners**
- **Concrete - Cracks or openings, signs of infiltration, crumbling, or rust**
- **Wood - Splitting, buckling or rotting**
- **Earthen containment walls - Settling, seepage, slumps, or animal burrows**
- **Wall alignment (vertical and horizontal) - curves or bulges**
- **Foundation - erosion or piping**
- **Underdrains - check that they are functioning as intended**

Leak Detection

In addition, every few years the facility should be cleaned so that an internal structural inspection by a qualified individual can be conducted. For lagoons that cannot be emptied, such as clay lined lagoons which should be kept moist to prevent the clay from drying and cracking, liquid balance tests may be performed. These tests monitor the liquid level in the lagoon. A leak is indicated if the liquid level drops more than can be accounted for by precipitation inputs and evaporative losses.

Monitoring Wells

If facilities cannot be emptied and inspected, it may be advisable to install groundwater-monitoring wells, e.g., clay-lined lagoons. Three monitoring wells are recommended -- one up-gradient and two down-gradient (relative to the direction of groundwater movement). Test wells at least annually for nitrate content and coliform bacteria.

Housekeeping and Aesthetics

Regular housekeeping is essential for efficiency, safety and public acceptance. Employees should clean equipment and grounds regularly, and collect and properly dispose of any trash generated; prevent it from blowing to adjacent sites. Sites that are visible from roads or adjacent properties should be regularly mowed and kept neat and clean.

Dust

Vehicle traffic is usually the primary source of dust at storage facilities. Speeds should be limited, and access lanes for larger facilities should be graveled. Dried Class A or Class B and composted materials may be dusty and require appropriate dust abatement in arid, windy climates, such as tarps. Care must be used to be sure the tarps are only used on heat dried biosolids that are

---

already very dry, and have not been rewetted, or compost that has been well stabilized or there may be re-heating. Tarps place on self-heating materials can enhance heat retention and contribute to spontaneous combustion of materials and fire.

Practices to Prevent Mud or Biosolids from being Tracked onto Public Roadways

1. Vehicles transporting biosolids should be cleaned before they leave the WWTP

2. Concrete or asphalt off-loading pads at the storage facility, will help keep equipment clean and make clean up of drips or spills easier.

3. The storage facility should have provisions to clean trucks and equipment when the need arises. Mud on tires or vehicles can be hand-scraped or removed with a high pressure washer or with compressed air (as long as this does not exacerbate an existing dust problem).

4. All vehicles should be inspected for cleanliness before leaving the site.

5. Use mud flaps on the back of dump trailers to preclude biosolids getting on tires or undercarriage during unloading operations.

6. Install a temporary gravel access pad as necessary at the entrance/exit to avoid soil ruts and tracking of mud onto roads.

7. Public roadways accessing the site should be inspected each day during operational periods, and cleaned promptly (shovel and sweep).

	Critical Control Point 3: Constructed Facilities Checklist	
	(Involving lagoons, pads, or storage tanks)	
	Project Management	✔
1	Prepare and maintain a Storage Site Management Plan with spill plan	
2	Critical Control Point 1: Work closely with the WWTP on stability and consistency	
3	Critical Control Point 2: Transportation; clearly mark site access routes and unloading areas	
4	Train employees to properly operate the storage facility and to perform inspections; conduct spill drills	
5	Maintain accurate and well organized records	
6	Designate a competent public relations person; maintain communications with stakeholders; notify agencies of reportable incidents; explain actions taken to respond to citizens concerns or complaints	
	Operations	✔
1	Minimize ponding and storage time; manage accumulated water properly	
2	Inspect and maintain up-and down-slope water diversion/collection systems	
3	Inspect and maintain tanks, ponds, curbs, gutters and sumps used to collect runoff	
4	Inspect buffer zones to ensure flow is not moving out of bounds	
5	Install signs and implement security measures to restrict public access	
6	Inspect concrete, wood, earth, walls, foundation and monitoring wells at constructed storage facilities	
7	Meet nutrient and hydraulic loading limits and state/local requirements when land applying accumulated water from storage	
8	Clean all vehicles and equipment before they exit onto a public road	
9	Train employees to use of appropriate sanitation practices; ensure practices are properly followed	
10	Retest nutrient and solids content prior to land application to re-calculate land application rate of biosolids, if the characteristics of the biosolids have changed significantly during storage	
11	Inspect for odors and conditions conducive to odors; mitigate appropriately	
12	Attend to site aesthetics	

Security

Lagoons, tanks, and some pads or bunkers for storage of liquid or dewatered biosolids are potential drowning hazards. When surface crusts form on the stored biosolids, they deceptively appear as though they will support a person's weight, but they will not. In addition, geotextile liners are generally smooth, and when wet, the sloping walls of lagoons may become so slippery that no foothold can be achieved. Facility perimeters should be posted with warning and no-trespassing signs. Fencing should be installed to keep out people and animals, and locking gates should be installed at vehicle access points. Appropriate rescue equipment such as life rings, lifelines, and poles should be kept on-site.

For aboveground tanks, ladders on the outside of tanks should terminate above the reach of people, or have locked barriers to restrict access to ladders; all access hatches should be locked. Personnel who access enclosed tanks must follow OSHA confined space entry guidelines and procedures, and have access to self-contained oxygen supply equipment when entering tanks.

IV. Odor Prevention and Mitigation

Prevention

Three key efforts to managing stored biosolids in a manner that prevents the development of odors include:

- **Only Store Properly Treated Biosolids**
  Ensure that only properly treated biosolids that meet all state and federal pathogen reduction regulations are delivered to the facility. Unless biosolids will be stored at remote sites for limited periods (60 days) and/or during cool weather months, vector attraction reduction should be met prior to storage.

- **Plan**: Develop written odor control and response plans.

- **Train**: Operator training can increase sensitivity of personnel to odor concerns and ensure proper implementation of the odor control plan.

- **Inspect, Monitor, Respond, and Record**: Regular inspections and odor monitoring, coupled with appropriate corrective action and recordkeeping, will help site and facility managers maintain good neighbor status and public acceptance of the project.

On an operational basis, use of the following management practices (where appropriate) may greatly reduce the potential for unacceptable off-site odors.

Practices to Reduce the Potential for Unacceptable Off-Site Odors

✓ Ensure that the WWTP has used processes that minimize odor during processing.

✓ Minimize storage time.

✓ Monitor and manage any water to prevent stagnant septic water accumulations.

✓ Avoid or minimize storage of biosolids during periods of hot and humid weather if possible. During warm weather, check for odors frequently. Use lime or other materials to control odors before they reach unacceptable levels off-site.

✓ Empty constructed storage facilities as soon as possible in the spring, for cleaning and inspection; keep idle until the following winter if possible.

✓ Select remote sites with generous buffers between sensitive neighbor areas.

✓ Consider weather conditions, prevailing wind directions, and the potential for off-site odors when scheduling and conducting clean-out/spreading operations. For example, operations on a hot, humid day, with an air inversion layer, and wind moving in the direction of a residential area on the day of the block party greatly increases the risk of odor complaints.

✓ Conduct loading/unloading and spreading operations as quickly and efficiently as possible to minimize the time that odors may be emitted. Surface crusts on stored biosolids seal in odors, but they break during handling, and odors can be released.

✓ Enclosed handling or pumping systems at constructed facilities may reduce the potential for odors on a day-to-day basis, but theses facilities still have the potential for odors during off-loading operations when active ventilation is used.

✓ Observe good housekeeping practices during facility loading and unloading. Clean trucks and equipment regularly to prevent biosolids build-up that may give rise to odors. If biosolids spills occur, clean up promptly.

Provide local government and state agency representatives with a contact name and number. Ask them to call the storage facility operator immediately if they receive citizen questions, concerns, or odor complaints resulting from storage of biosolids. Operator staff should politely receive citizen questions or complaints, collect the individual's name and phone number, conduct a prompt investigation, undertake control measures, if necessary, follow-up with the person who filed the complaint, and document the event and actions.

Mitigation

If significant odor should develop during handling operations, the following remedial measures can be taken:

Odor Remediation Measures for Use During Handling Operations

✓ Immediately correct any poor housekeeping problems (such as dirty equipment).

✓ Immediately treat any accumulated water that has turned septic with lime, chlorine, potassium permanganate or other odor control product; remove the water as quickly as possible to a suitable land application site.

✓ If odors are arising from lime stabilized biosolids, pH should be measured. If it has dropped below 9.0, lime can be applied, topically to dewatered material, or, in highly liquid systems, lime slurry can be blended into the biosolids by circulation. The pH should be monitored and dosed with lime until the desired pH has been achieved. Raising pH halts organic matter decomposition in the biosolids that can generate odorous compounds.

✓ For most types of biosolids (digested, lime stabilized, liquid, dewatered), applying a topical lime slurry will raise surface pH levels, create a crust, and reduce odors. Topical spray applications of potassium permanganate ($KMnO_4$) or enzymatic odor control products to neutralize odorous compounds may also be effective in some situations.

✓ Cover biosolids with compost or sawdust.

✓ If the odor is due to the combination of wind and weather conditions (hot, humid) and agitation and circulation of biosolids as part of unloading operations, it may be advisable to cease unloading operations until weather conditions are less likely to transport odors to sensitive off-site receptors.

✓ Spread and incorporate or inject odorous material as quickly as possible.

✓ For enclosed storage facilities, absorptive devices (charcoal or biofilters) incorporated into a ventilation system may be a feasible option for reducing odorous emissions.

✓ Cause the WWTP to change its processes to produce less odorous biosolids.

V. Spill Prevention and Response

Prevention

Liquid tankers, and trailers used for semisolid biosolids, should have rubber seals around all hatches and tailgates that can be mechanically tightened to prevent any leakage. At the beginning of each day, inspect the seal integrity on all vehicles. After loading, check each unit for leakage prior to operating the unit on public roadways. Seepage or dripping of biosolids is unacceptable.

When liquid biosolids are being handled, it is recommended that buckets be placed under hose connections to collect any drips when hoses are connected and disconnected. In addition, paving and curbing of the off-loading pad facilitates collection of small quantities of biosolids that may drip or spill.

Spill Response

A spill response plan should be a special part of the site management plan. Examples of the spill response plan and accompanying biosolids fact sheets used by Los Angles County Sanitation District are shown at the end of this chapter. Furthermore, staff should be trained to follow the plan. This means conducting periodic training and 'spill drills' that include training on contact with the media.

To ensure prompt reporting and initiation of clean-up activities, it is recommended that site supervisors have access to cell phones or to two-way radios. Also, road tractors and application equipment should have cell phones. If a spill occurs, the site supervisor should immediately initiate clean-up. The site supervisor should also contact appropriate emergency services if necessary (i.e. fire or rescue); notify supervisors; and communicate with the public on the scene or notify the designated community contact, and appropriate state regulatory agency. Site workers should also have media contact training.

The first step in the clean-up process is to ensure public and worker safety. Next, halt the source of the spill, e.g., a ruptured line or valve or damaged tanker unit, and contain the spill. In the event large quantities of liquid or semi-liquid biosolids are spilled off-site, straw bales, where available, may be used to contain and soak up biosolids.

Once the source is controlled, collect spilled material. For liquid spills, vacuum equipment on biosolids application vehicles can be used to collect as much material as possible. Residual amounts are usually removed by hand shoveling or sweeping. Straw, cat litter, or commercial adsorbents may be spread as necessary to complete removal of the material. Absorbent materials should be swept or shoveled up and taken to a permitted land application site or to an approved landfill. If necessary, roadways may then be flushed with water to complete the clean-up process.

Reporting

Prior to initiating a field storage operation, it may be advisable to contact the local police, fire, and hospital teams to brief them on the facility and its operation, including risks and types of injury that could potentially occur at the site. In the event of a spill or leak, state and local regulators with oversight responsibilities for the facility should be notified as required by state and local regulations. Generally, a written report documenting how the spill occurred and all remedial actions should be completed promptly after the incident and submitted to the regulatory authority or kept on file.

Biosolids Fact Sheet[1]
(Generator/ facility name)

DESCRIPTION

Biosolids (formerly referred to as sewage sludge) are reusable solids from the wastewater treatment process. At_____(treatment plant name), biosolids have been treated by _____(process, e.g. , anaerobic digestion) and dewatered by _____ (process type , e.g., filter presses). The dewatered, semi-solid form is referred to as cake.

Biosolids are not a hazardous material. The biosolids cake produced at _____ _____(treatment plant name) is primarily organic. It is beneficially reused as a soil amendment on agricultural land (land application), _____ (other uses here, e.g., compost). Routine analyses demonstrate that _____ (quality/allowable use, e.g., metals concentrations) meet EPA standards that allow the material to be land applied at unrestricted metals loading rates.

(Further information here, e.g., anaerobic digestion significantly reduces, but does not completely eliminate, pathogens (disease causing microorganisms). Digesters, which are operated at specific time and temperature parameters, produce EPA Class B biosolids. Class B quality is suitable for application to agricultural land in concert with certain EPA site restrictions.)

TYPICAL CHARACTERIZATION

Appearance	**Black, semi-solid**
Total Solids Content	_____ % (_____ % moisture)
Free Liquid	**None**
pH	_____
Nitrogen	_____ % (dry weight basis)
Phospate	_____ % (dry weight basis)
Potassium	_____ % (dry weight basis)
Metals Content	_____e.g., Meets EPA Table 3
Pathogen Reduction	_____e.g., Meets EPA Class B
Soluble Metals	_____e.g., Non-hazardous per_____ STLC and
TTLC	

(State)

HANDLING PRACTICES[2]

Biosolids are treated to reduce pathogens. Nonetheless, there is the potential for exposure to pathogenic microorganisms. Major routes of infection are ingestion, inhalation and direct contact. Good, common sense, personal hygiene and work habits provide adequate protection for workers handling biosolids. Recommendations include:

---

[1]Fact sheet was provided courtesy of Los Angeles County Sanitation District
[2]Much of the information contained herein was taken from Biological Hazards at Wastewater Treatment Facilities, Water Environment Federation (formerly, Water Pollution Control Federation), 1991.

- Always wash hands after contact with biosolids.

- Avoid touching face, mouth, eyes, nose, or genitalia before washing hands.

- Eat in designated areas away from biosolids handling activities.

- Do not smoke or chew tobacco or gum while working in direct contact with biosolids

- Use gloves, when applicable.

- Keep wounds covered with clean, dry bandages.

- Change into clean work clothing on a daily basis.

- If contact occurs, wash contact area thoroughly with soap and water. Use antiseptic solutions on wounds, and bandage with a clean, dry dressing. For contact with eyes, flush thoroughly but gently.

- The Centers for Disease Control recommends that immunizations for diphtheria and tetanus be current for the general public. Boosters are recommended every ten years. The tetanus booster should be repeated in the case of a wound that becomes dirty if the previous booster is over five years old. Consult a doctor regarding direct exposure to an open wound or mouth.

HAZARD POTENTIAL

Biosolids are not combustible under ordinary circumstances. If stored in airtight containers for an extended period, methane gas may be produced which could ignite in the presence of a spark or open flame. Extinguish with dry chemical, water spray or foam. Avoid use of open flames in confined areas and around sealed transport containers. Vent confined areas and transport containers if biosolids have been stored for any significant length of time.

Hydrogen sulfide may also be generated in sufficient quantities to be a hazard in enclosed areas such as tarpped transport containers. Hydrogen sulfide gas, which smells like rotten eggs, can be toxic. Exposure can be avoided by removing the container tarp prior to unloading, and discharging as much material as possible prior to employees entering the container.

GENERATOR DATA

Generator Name | Facility Name (if different)
Address | Address
City, State, Zip Code | City, State, Zip Code
Area Code & Phone Number | Area Code &Phone Number
Contact | Contact

Biosolids Hauler Spill Response Procedure[3]

1. **General**

A. Biosolids are non-hazardous and non-toxic. If a spill occurs, there is no need for special equipment or emergency protocol beyond that outlined in this procedure. Biosolids are primarily processed solids produced by sewage treatment plants.

B. Biosolids spilled onto pavement pose a potential road hazard because they can create wet, slick surfaces for motor vehicles, and/or can obstruct traffic flow. If biosolids remain on the surface for a sufficient time, they could be a source of potential contamination of nearby storm drains, waterways, or ground water. Biosolids should be thoroughly removed so that no significant residues remain to be washed into any storm drain or waterway by surface water. All spilled biosolids must be returned to the trailer from which they spilled, or be loaded into another appropriate transport vehicle.

2. **Biosolids Characteristics and Personal Hygiene Procedures**

A. Biosolids are processed organic residual solids from domestic sewage treatment, containing nitrogen, phosphorus, trace metals, and some pathogenic (disease-causing) organisms. Biosolids being transported are typically _____ % total solids, with a _____consistency (Fill in description). Biosolids become dirt-like when solids exceed 45%. The material contains x % volatile solids, with a pH of

    .

B. Personnel cleaning up a spill of biosolids should:

    • Wear gloves for shoveling, sweeping or handling biosolids.

    • Not eat, drink, smoke or chew while working directly with biosolids

    • Wash hands (and as necessary all other exposed parts of the body) with waterless hand cleaner, or soap and water, following spill clean-up and prior to eating, drinking, smoking or chewing.

3. **Over-the-Road Spill Response Procedures**

A. Park the truck on the side of the road and place traffic cones, reflectors and/or flares to divert traffic around the spill. Remain with the truck and spilled materials, unless it is necessary to leave temporarily to contact emergency services.

B. Drivers shall notify their Supervisor as soon as possible by radio or by phone (Area code & phone number) _____. Give the location and amount of biosolids spilled. Also notify the California Highway Patrol by telephone [911], if the spill has occurred on a public right of way.

---

[3] *Procedure courtesy of Los Angeles County Sanitation District*

C.   Inform the authorities that you are hauling biosolids which is non-hazardous and non-toxic.

D.   Cooperate with the authorities, assist with traffic control and clean-up.

E.   Do not leave the scene of any spill, even a small one, until it is cleaned up. You may clean up small spills first and then report the spill.

4.   Spill Response Procedures

A.   Load spilled biosolids back into the vehicle if it is operable. If the vehicle is disabled, the spill must be loaded into an alternate vehicle.

B.   Spilled biosolids must be prevented from migrating off the incident site, into storm drains, or into surface waters. This is especially important if an incident occurs in rain conditions. Biosolids spills may be diked or controlled with sand, sand bags, straw, absorbents, or other blocking material.

C.   Two people working with shovels can load a small spill into a vehicle. A large spill must be loaded into the vehicle by an appropriate rubber tired loader. The scene coordinator is best suited to choose the appropriate loading option to deal with the spill, based on equipment availability and spill size.

D.   After the spill has been loaded, the incident site must be cleaned. Spills may be cleaned by sweeping the site free of remaining debris. Do not wash off tools or trucks at the spill location; return tools and trucks to the wastewater treatment plant for cleaning.

E.   Cleaned up spills should either be taken to the original destination or to a landfill permitted to receive biosolids. They may also be accepted by the originating sewage treatment plan.

F.   Spill response drills should be conducted periodically.

References

Brinton, W.F., Jr., E. Evans, M.L. Droffner, and R.B. Brinton. 1995 Standardized test for evaluation of compost self heating. BioCycle 36 (11): 64-69.

National Fire Protection Association. 1994. NFPA 298: Standard on Fire Fighting Foam, Chemicals for Class A Fuels in Rural, Suburban, and Vegetated Areas. NFPA, Quincy, MA

National Fire Protection Association. 1997. NFPA 299: Standard for Protection of Life and Property from Wildfire. NFPA, Quincy, MA

NRCS. 1992. Agricultural Waste Management Field Handbook, Part 651. National Engineering Handbook 210-VI. Natural Resource Conservation Service, USDA.

Sullivan, D.M., D.M. Granatstein, C.G. Cogger, C.L. Henry, and K/P. Dorsey. 1993. Biosolids management guidlines for Washington State. Washington State Dept. of Ecology Publication 93-80.

Sullivan, D.M. 1999. Towaqrd Quality Biosolids Management: A Trainer's Manual. Northwest Biosolids Management Assoc. Seattle, WA.

USDA. National Handbook of Conservation Practices. Natural Resource Conservation Service. http://www.ftw.nrcs.usda.gov/nhcp_2.html for specific engineering and practice standards about Diversion (362), Composting Facility (317), Field Border (386), Filter Strip (393), Hillside Ditch (423), Runoff Management System (570), Waste Management System (312), Waste Storage Facility (313), Waste Treatment Lagoon (359), Waste Utilization (633).

USDA. 1992. Agricultural Waste Management Field Handbook, Part 651. National Engineering Handbook 210-VI. Natural Resource Conservation Service. Washington, D.C.

Wilber, C. (ed.) 2000. Operations and Design at the Wastewater Treatment Plant to Control Ultimate Recycling and Disposal Odors of Biosolids. USEPA sponsored project.

Chapter 6

Community Relations

Introduction

Whether a biosolids storage site is located in a remote area or in one that is more densely populated, developing a relationship between project proponents and the surrounding community is critical to successful field storage. The public's view of the benefits of biosolids recycling and the necessity for biosolids storage, as part of well-run land application programs, frequently are balanced by concerns regarding potential environmental, health or nuisance impacts. Issues commonly raised about storage sites include potential odors, noise, dust, traffic, human or animal health effects, and water quality or environmental impairment. These concerns are often linked to broader issues such as potential impacts on property values, compatibility with other land uses, and political issues. For these reasons, biosolids field storage projects, either in small field stockpiles or in large, permanently constructed facilities, should include a community relations program. The relationship that the storer/applier develops with the community is just as, or more important than, the one between the biosolids generator and the applier. Table 6-1 identifies potential issues and community concerns related to field storage of biosolids.

Table 6-1. Common Issues and Community Concerns about Field Storage of Biosolids

Issue	Community Concerns
Air Quality	Odors, dust and pathogens
Water Quality	Surface runoff to streams and well water contamination with respect to nutrients, toxic metals, organics and pathogens
Public and Animal Health	Contact and potential disease transmission, inadequate buffer zones, and animal grazing
Traffic and Safety	Posting and access control, road conditions and speeding
Aesthetics	Odors; visibility, noise, dust and property values

The ultimate goal of the community relations program is to develop public acceptance of biosolids storage within the community. The size and extent of the community relations program depends on public interest more so than on project size. In general, large, capital intensive, constructed storage facilities, and facilities in high population areas, will require the greatest community relations effort. It is not uncommon for large constructed facilities at remote sites to attract less public interest than smaller highly visible projects.

Extensive education and outreach programs are most efficiently conducted on an ongoing basis, in the context of an entire biosolids recycling program, not just the storage component. Communications efforts related to storage issues would be most appropriately handled by being integrated into ongoing community relations efforts conducted by biosolids managers and WWTPs. Outreach programs should be initiated as early as possible, when biosolids projects are in the initial planning phase. The public desires a voice in activities that may impact their community, and they need to know that biosolids managers share their concerns and are responsive to their comments. Seeking early input from local officials and the citizens during the planning phase is the best way to gain public support. Active listening and responsiveness to public concerns builds trust and ensures that the project fits successfully into the community.

Communications programs should present all the pros and cons of a proposed storage site relative to its role in the land application program. Risks should be explained in terms that are understandable to the public. Biosolids generators, storers and appliers must be able to provide concrete answers in response to questions and concerns. Before a community can be involved, it must be informed and invited to participate. The basic communications elements that should be implemented prior to the initiation of any biosolids storage activities, especially long-term constructed facilities, are as follows:

1. At the inception of the project, arrange to brief local officials and staff (i.e. county supervisors, planning and zoning staff, Extension Service and soil conservation district staff) one-to-one on your plans. Solicit their input on suitable sites and potential local concerns.

2. Inform adjacent property owners and the local community particularly for constructed facilities. This may be accomplished through informal contacts and/or as part of formal notices and meetings or hearings associated with state or local permitting requirements.

3. Look for ways to adapt your project to accommodate legitimate local concerns. Be prepared to address the pros and cons of the project and hand out fact sheets answering the most frequently asked questions. Invite local officials and concerned citizens to tour existing field stockpiles or constructed facilities.

4. Develop a plan to promptly and effectively address public questions or complaints on an ongoing basis once the site is in operation. Be sure people know how to get in touch with you and maintain open channels of communication and feedback throughout the life of the project.

Audience Assessment

Managers of biosolids projects should consider that the "public" is not one homogenous group. Community relations efforts will be more effective if education and outreach efforts are targeted and tailored to address the particular concerns and interests of specific groups within the community. Key subgroups frequently involved in siting and operation of field storage areas are:

Elected Officials/Local Government Agencies
These individuals and organizations may have a regulatory role in the siting and development of storage facilities. They may have a role in selecting biosolids management options for their community. Elected officials in particular will want to ensure that the concerns of their constituencies are addressed.

Citizens Groups
Established organizations in the community (e.g., Rotary Clubs, the Chamber of Commerce, League of Women Voters) as well as ad-hoc groups established in response to the proposed project may be interested in storage projects. Their concerns may focus on the potential impacts of the biosolids activity on the immediate community, and include a wide range of topics (e.g., economic development, property values, agricultural and open space preservation, traffic impacts, aesthetics and health and environmental protection).

Agricultural Organizations
Organizations such as the Farm Bureau, USDA Cooperative Extension Service, Natural Resources Conservation Service and local Soil Conservation districts frequently take an interest in biosolids storage and land application programs from the perspective of providing economic benefits to farmers and landowners, and ensuring long-term protection and improvement of soil and water resources. In addition, organizations such as local conservation districts are excellent sources of technical information to assist in appropriate site selection and project development. Their participation in the project will help assure that local concerns are addressed.

Environmental Organizations
National environmental groups with local chapters and groups dedicated to local and regional environmental issues may take an interest in biosolids storage and use projects. Their focus may be related to water quality, environmental protection and improvement; recycling, or land use and development issues.

Local Media
Local media includes newspapers, television and radio stations that generally focus on public discussion on such issues.

Biosolids Users
Members of the local community who have personal experience using or storing biosolids on their properties should be requested to share their

---

perspectives on the pros and cons involved. Generally people who are known and respected in the community are a key source of information.

Employees

Employees (contracting agency and biosolids land applier/storer), particularly those that reside in the local community, are also a valuable part of community relations efforts. Employees should be briefed on the project. They can share information on the project through their informal contacts in the community, help ensure that public inquiries are promptly referred to the appropriate individual in the organization, or serve as representatives to area-wide planning groups, technical advisory committees or other community organizations.

Working successfully with diverse community groups may take special communication and mediation knowledge, training and experience. Assistance from a public relations professional may be needed.

Educational Tools

Once various audiences and issues are identified, there are a number of mechanisms that can be used to effectively disseminate educational materials and open lines of communication and participation. The following is a list of the most commonly used methods and pointers for using them effectively:

One to One meetings

The most effective community relations tool is usually one-to-one personal contacts. Identifying key individuals in the community and spending the time to meet with them personally is the best way to disseminate information, gain credibility and ensure that local concerns are identified and addressed.

News and media coverage

Publish meeting dates, times and locations. Invite the press to public meetings, tours or field days. Provide briefing packages on the project and contacts. Provide interviews or issue news releases.

Newsletters

Broad circulation of educational information can be achieved by contacting local organizations and asking them to feature an article you have prepared concerning the proposed biosolids project (e.g., agricultural extension service, chamber of commerce, environmental groups).

Fact Sheets/Displays

Develop fact sheets and displays for use at public meetings, libraries, and local events.

Public Meetings/Hearings

Offer to make presentations about biosolids at meetings of various groups. If public interest or regulatory requirements mandate it, conduct public meetings or hearings specifically concerning the proposed storage project.

Presentations to Schools/Youth Activities

Presentations to schools directly increase students' level of knowledge, and may result in second hand education of parents as well. Sponsoring student activities is a gesture of community support, and may provide another venue for disseminating project information to the public.

Tours/Field Days

Participate in local agricultural field days through on-site demonstrations, presentations, or exhibits. Organize educational tours of biosolids storage and land application sites for specific groups (e.g., local reporters, elected officials, community or environmental organizations).

Community Advisory Committees

Assemble a community or technical advisory committee. This type of community involvement is generally limited to situations involving permanent constructed storage facilities. Frequently such committees will be formed at the request of the local government. Committees of this nature take a more active role in the planning and design or storage facilities, management and operational plans and project oversight.

Program Evaluation

The success or effectiveness of a community relations program can be evaluated based on some of the following:

- Requests for information
- The tone of news articles and media coverage
- Endorsement from various organizations
- Absence of organized opposition to the facility and the continued operation of storage and land application activities.

It is important that once a program is through the initial planning stages, that on-going contact and communication is maintained in order to obtain regular feedback and address any local issues that arise promptly and effectively.

Chapter 7

Biosolids-Derived By-Products and Other Organic Materials

Introduction

The management practices recommended in this biosolids field storage guidance document are also generally applicable to storage of other types of non-hazardous organic residuals that are suitable for recycling and beneficial use as a fertilizer or soil conditioner. These materials may be used for agricultural, horticultural, reclamation, landscaping or landfill cover purposes. Storage is frequently desirable for these products due to seasonal markets for some materials (e.g., compost or topsoil), crop cycles, and weather restraints on land application programs. Organic residuals may be generated through industrial or agricultural processes and include biosolids-derived products that serve as topsoil. Examples of these materials are provided below. A more extensive list of organic materials is provided in Appendix E.

Other Organic By-Products

Biosolids blended topsoil
Yardwaste (leaves, grass clippings, woodchips)
Food processing residuals (fruit and vegetable peelings, pulp, pits)
Meat, seafood, poultry and dairy processing wastewater and solids
Hatchery wastes
Animal manure and bedding
Waste grain, silage
Spent mushroom substrate
Wood ash
Pharmaceutical and brewery waste
Pulp and paper mill residues
Mixed refuse (food scraps, paper etc.)
Textile residuals

Storage Considerations

Some organic residuals are unmodified (e.g., vegetable peelings, wood ash, etc.), others are generated through wastewater treatment processes (slaughterhouse wastes), or undergo composting, blending, or other treatment methods. The physical consistency of these residual materials, may be either liquid, semi-solid/dewatered, or dry.

As with biosolids, locating suitable sites, and the development and implementation of practices to deal with storage and handling of these materials will benefit from considering the Critical Control Points approach described in Chapter 1 to address odors, water quality, pathogens, field management practices and community relations. Depending on the material in question, some of these issues may be more significant than others. To determine which combination of management practices, handling techniques, and storage options is most suitable, the following specific product characteristics should be evaluated:

Physical consistency and water content

Biological Stability
Pathogen Potential
Odor Characteristics
Vector Attraction
Nutrient and BOD Content
Fats and Oils
Dust Potential
Combustibility
Consistency and predictability of product

Physical consistency and water content

The physical consistency and solids content of the material, whether liquid, semisolid, or dewatered or dried, is essential for evaluating the suitability of the material for various types of storage options and is essential for planning storage capacities. Generally, materials with solids contents less than 12 percent are not appropriate for field stockpiles because the material is too wet to hold shape and will slump and flow. Storage of these materials is best accomplished in lagoons, tanks, or basins. However, dried, composted, dewatered materials may be suitable for either field stockpiles or constructed storage facilities.

The percent solids in liquid and semisolid materials may change over time due to precipitation or evaporative losses. Solids may also settle during storage. Depending on the degree of liquid/solids separation and the amount of recirculation possible to resuspend solids prior to removal, it may be necessary to retest nitrogen and percent solids to determine appropriate application rates.

Biological Stability

Some organic residuals contain organic constituents that are easily digestible (decomposable) by microorganisms and others do not. Materials not biologically stabilized through composting or other treatments (Table 2.1), will require a higher level of management during storage to prevent the development of unacceptable odors or attraction of flies or other nuisance vectors. Other organic residuals, that are not easily digestible, present minimal potential for the generation of nuisance odors. In some cases, storage also allows blended ingredients to react further with each other (as in curing or aging phase with compost) and this produces a more stable material with less odor potential when it is ultimately land applied.

Consideration of the biological stability of the material to be stored is a key factor in siting decisions (such as suitable buffers) and in selection of appropriate storage methods and management practices. Explaining how the operations methods and practices are suited to deal with the type of biosolids and its degrees of biological stability is an additional and important way to gain community acceptance.

Pathogen Potential

Certain organic residuals (such as poultry processing wastes or animal manures) may contain pathogens at levels similar to or higher than the limits established for Class B biosolids. These materials can have a potentially negative impact on human or animal health if they are not properly managed.

In some instances, these materials may be disinfected or stabilized prior to storage, or the storage period itself can provide time for pathogen die off. If the material is a biosolids blend that must meet Class A standards, testing for pathogens as per the Part 503 regulations testing will be necessary.

Odor Characteristics

Offensive odors in most organic residuals are generated during microbial decomposition of the organic matter constituents. In some instances, a material contains residual levels of compounds that are inherently odorous but do not result from biological decomposition. The potential for release of unacceptable levels of odorous compounds is most likely when materials are agitated, mixed, or moved. Stabilization processes (Table 2.1) used to control pathogens generally also help reduce potential odor levels. Other methods or management practices for odor management include: moisture reduction, maintenance of aerobic conditions, pH adjustment, enclosed handling and storage, cold weather storage, minimization of storage duration during hot humid weather, and keeping dried materials dry in the field. A useful technique to reduce odor from stored materials is to cover them with compost or sawdust. Field storage of highly odorous materials may require either remote sites or enclosed handling systems (e.g. tanks and subsurface injection applicators). Depending on the stability of the product and storage conditions, the potential for off-site odor may increase the longer (months or more) the material is stored and the greater the volume of stored material.

Vector Attraction

Organic residuals such as food processing wastes and animal manures, or other unstabilized materials may be attractive to flies or vermin, which can create nuisance conditions or, with certain materials, are a potential pathway for pathogen transmission. To foster community acceptance, materials must be managed in a manner that controls vectors and prevents off-site nuisances.

Nutrient and BOD Content

The nitrogen content and its form in organic residuals depends on the type of material, handling and storage methods, and the length of storage. Materials with high ammonia levels can easily lose this nutrient through volatilization. Appropriate handling and storage options can reduce odor potential and conserve this plant nutrient.

High biological oxygen demand (BOD) reflects the readily degradable organic matter in the material. Many untreated organic residuals, particularly those containing oils and greases, have a high BOD. This means that the material is subject to microbial decomposition and possibly to anaerobic conditions that may generate odors during storage. Materials with higher nutrient levels and BOD also have a greater potential to impact water quality if they escape to waterways.

The longer organic materials are stored, the greater the potential for the nutrient content, total solids, and salt content or pH to change. With some materials, testing for these parameters before removal may be advisable to properly calculate land application rates.

Fats and Oils

Materials that contain significant amounts of fats and oil (e.g. meat processing wastes, grease trap wastes) can be highly odorous. Significant management is required to prevent unacceptable odor levels at storage sites. Remote site locations for open-air storage may be sufficient in some cases, but in many localities, enclosed handling using pumps, hoses and tanks may be necessary to control odors. Fats and oils also contribute to high BOD. These materials may also present handling challenges caused by clogging or gumming up of equipment.

Dust Potential

Dried residuals such as composts and wood ash may generate dust during dry windy conditions. The potential of a material to create dust should be kept in mind during site selection and these materials must be managed to alleviate off-site nuisance conditions.

Combustibility

Immature composts, wood chips and yard waste, poultry litter, biosolids blends, or heat dried materials may be combustible and/or, under certain conditions, undergo self-heating and spontaneous combustion from the heat generated by microbial decomposition. Wetting of dry material or confined storage, which traps heat, may exacerbate these conditions. Management plans should be developed to prevent this occurrence and contingency plans should be in place to respond appropriately if self-heating occurs.

Consistency/Predictability of Product Over Time

Consistency of the product's characteristics over time and the volume or amount produced over the course of a year should be considered. Certain facilities may produce greater quantities of an organic residual at certain times of the year (e.g., yardwaste) or the product characteristics may change over the course of a year (e.g. vegetable wastes at a cannery change as different crops are harvested and processed). The variability of a material in terms of volume or product characteristics may require increased flexibility in management and closer coordination of the storage and land application components.

Regulatory Considerations

Federal and state regulations governing organic residuals vary with the type or origin of the material, so the applicable laws for any given material must be investigated (see Appendix F for a list of state agency contacts). The land application of certain organic residuals is regulated under 40 CFR 257 "Criteria for Classification of Solid Waste Disposal Facilities and Practices" under the Resource Conservation and Recovery Act. However, theses criteria do not apply to agricultural wastes, including manures and crop residues. The Federal Part 257 regulations do not address storage issues specifically, but this regulation does include provisions regarding general management of these materials. For instance, residuals management practices conducted in floodplains may not restrict the flow of the base flood, reduce temporary water storage capacity of the floodplain, or result in washout of solid waste, so as to pose a hazard to human life, wildlife, or land or water resources. Likewise, practices may not: impact threatened or endangered species or habitat; be either a direct discharge or a nonpoint source of pollutants; or contaminate underground drinking water sources. In addition, Part 257 requires control of on-site populations of disease vectors.

From state to state, the degree of regulation governing the handling, transportation, storage, and beneficial use of organic residual materials varies widely. Some states require permits for land application or storage of these materials, similar to those for biosolids. Other states do not have comprehensive regulations or permitting requirements for all, or some types of, these materials. Therefore, it is important that residual managers investigate the regulations thoroughly prior to initiating a storage and land application program. In addition, if a constructed storage facility is proposed, local zoning and building permit requirements will need to be investigated.

References

Brandt, R. C. and K. S. Martin. 1994, The Food Processing Residual Management Manual Pennsylvania Department of Environmental Resources, Harrisburg, PA. Pub. No. 2500-BK-DER-1649.

Appendix A

Odor Characterization, Assessment and Sampling

Odor Characteristics

Odors are characterized and measured by their psycho-sensory, social, and somatic impacts as well as by their physical-chemical properties.

Sensory Characterization

Sensory evaluation of odors involves description of the odor character as well as measurement of odor intensity, pervasiveness, and quantity. **Character** of an odor is a word description of what it smells like, e.g., rotten cabbage, rose, cinnamon. The character of an odor and its desirability (good, bad, or neutral) influences its acceptability when perceived.

Intensity is a measure of the perceived strength of an odor. This is determined by comparing the odorous sample with a "standard" odor, often various concentrations of n-butanol in odor-free air. Intensity is expressed in terms of micrograms per liter of butanol (µg/l) in liquid, milligram per cu. meter (mg/cu. m) in air, or ppm butanol. Intensity is also used to calculate pervasiveness.

Pervasiveness (persistence) describes how noticeable or detectable an odorant is as it's concentration changes. A pervasive odor is one that can be perceived by people even though the odor has been diluted many times. Pervasiveness of an odor is determined by serially diluting the odorant-containing sample and measuring the intensity at each dilution. When the results are plotted on log-log paper, an intensity slope is established. A flat slope (e.g., 0.2) would reflect a very pervasive odor because the odor can still be detected after millions of dilutions. Conversely, a steeper slope (e.g., 0.5) would reflect a much less pervasive odor, or one that would not be detectable after only a few hundred dilutions. Organic sulfur-containing compounds, e.g., dimethyl disulfide, can often be described as pervasive because the odor may be detected off-site where it is present at very low concentration. The fact that it is not smelled on-site even though it is present at higher concentrations than it is off-site, can be explained by the masking effects of ammonia. The latter typically would have such an intense odor close to the source, that other co-occurring odorants would not be perceived.

Quantity of odor, as measured on a sensory response scale (i.e., based on odor detection), is expressed in terms of how many dilutions it takes before it is no longer detectable, although the exact character of the odor may not be discernible. This is often expressed as dilutions to threshold or odor units.

If the quantity is expressed in parts per million (ppm) or billion (ppb) or in moles or micrograms per cubic meter of specific chemical compound, then the determination is no longer sensory, rather, the value represents the physical, chemical amount of an odorant (explained in greater detail in this appendix).

Odor Assessment

Effective management of odorous emissions requires a systematic method for odor assessment and sampling. This can involve a perceptual response method, an analytical instrument approach, or a process that uses elements of both approaches. Regardless of how specific odorants are determined (chemically or perceptually), managing odorous emissions and alleviating odor nuisance remains the desired end result of odor evaluations and assessments.

Field Practice Options

Several approaches available for field assessments of odor include:

1. pro-active use of on-site and community odor surveys by site or facility operator and staff (see the Springfield odor survey forms at the end of this appendix)

2. use of portable sensory instruments by trained odor inspectors (see the St. Croix sensory example performance standard procedure at the end of this appendix)

3. application of public nuisance criteria

4. evaluation of odor samples by an odor panel

5. use of an annoyance survey coupled with quantitative chemical analysis of odorous air samples in a potentially impacted community

6. establishment of quantitative standards for known odorous compounds coupled with regular air sampling and chemical analysis

Although several of these approaches (1,2,4) use measurement and evaluation, they may fail to provide accurate assessments for several reasons. First, the concentration of the offending compound(s) may be below current standards. Second, there may not be standards for them, or, third, in the case of the odor panel, responses may not correspond to the evaluations of people in the affected community. For these reasons, odor or annoyance surveys (approach 5) may assist operators, communities, and regulators in fairly determining and evaluating odor problems and effectiveness of abatement actions.

The use of odor or annoyance surveys, especially in combination with air sampling (approach 6), can help objectively determine the presence or absence of nuisance odors in a community. This approach differs significantly from the three typical approaches used by regulatory agencies to deal with odor problems. In addition to collection of air samples for odorous compounds in an affected community (such as described below), an odor or annoyance assessment might include a scientifically designed public opinion survey, which draws opinions from <u>randomly</u> selected individuals in the community. To keep the odor component of a community survey unbiased relative to other community annoyances and environmental impacts, the survey may also include questions about other environmental factors such as noise, traffic, stray or wild animals, and other community characteristics.

Physical-Chemical

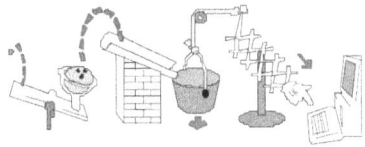

Both organic and inorganic compounds have been identified as odorous constituents of wastewater, solids, and biosolids. Compounds typically of concern can be formed during aerobic or anaerobic decomposition of proteins and carbohydrates that are abundant in wastewater and biosolids. Table A-2 lists common odorous compounds associated with biosolids. Many of these compounds are intense (see discussion below) and have odor thresholds in the parts per billion (ppb) concentration ranges. Odor threshold is the minimum concentration required for an individual to perceive the odorant. The main odorants emitted from biosolids include:

Ammonia. Ammonia is most often found in emissions from freshly alkaline stabilized materials and during early phase composting. Table A-1 shows the considerably greater odor threshold for ammonia than for reduced sulfur compounds. At least 100 to 1000 times more ammonia than reduced sulfur compound is needed per unit volume of air for an average person to detect it, even with the variation in reported odor thresholds.

Ammonia also has an important special characteristic that field site operators need to recognize. At high concentrations, it is so intense that it strongly masks odors from other compounds, such as those containing reduced sulfur groups. Thus, a misleading assessment report indicating no potential for off-site odor, could result if only ammonia were detected directly at the field storage site. In fact, reduced sulfur compounds also might be present, but not detectable, because of ammonia masking. However, as the air 'parcel' containing both types of compounds moves downwind, beyond the storage/application site perimeter, ammonia could be diluted below its detection threshold. In contrast, the reduced sulfur compounds, although also diluted below their on-site concentrations, may still be concentrated enough to remain above their detection thresholds. For this reason, odor assessments at field storage sites should include some monitoring for off-site reduced sulfur or amine odors.

Ammonia that is emitted comes from anaerobic bacterial digestion of proteins found in the stored materials. As the pH of the materials increases above 8.0, more ammonia is released. Ammonia is often accompanied by release of amines, and if chlorine is used, chloramines may be released as well.

Inorganic sulfur compounds such as hydrogen sulfide. Hydrogen sulfide (H_2S) often gets the most attention because of the familiar rotten egg odor associated with it. However, it is rarely detected in field stockpiles. Often other compounds or combinations of compounds listed in Table A-2 are the primary cause of odor in biosolids. When pH is less than 9.0, hydrogen sulfide can be generated from wastewater solids under anaerobic conditions. Increasing the pH to 9.0 or higher, as happens when biosolids are lime stabilized, can eliminate H_2S emissions.

Organic sulfur compounds. Dimethyl disulfide (DMDS) and dimethyl sulfide have been associated with odorous emissions from biosolids composting operations. Also, it has been measured at wastewater solids and dewatering facilities, pelletizing facilities, and digester gas. In general, DMDS is a by-product of chemical or microbial degradation (anaerobic) of proteins.

Mercaptans or thiols are a generic class of straight-chained organic compounds that contain a single sulfur molecule. Methyl mercaptan is the most common thiol measured in biosolids emissions. Table A-2 shows methyl mercaptan has a low odor detection threshold, i.e., quite small amounts are easily detectable. Thus, its presence can lead to odor complaints. Two methyl mercaptan molecules combine to form one DMDS molecule. Active ingredients of garlic (allyl sulfide) and onions (propanethiol) have precursors that are similar to mercaptans; spoiled broccoli also produces mercaptans and DMDS. The boiling point of methyl mercaptan is 6°C, which makes it a gas at room temperature. Therefore, measurement techniques that use tedlar bags are acceptable.

Volatile fatty acids (VFAs). These short chain ($< C_8$) fatty acids have the general formula C_nH_{2n} +COOH and are typically generated during anaerobic decomposition of vegetable materials, such as hay, straw, grass, leaves, silage, etc. VFAs include: formic, acetic, propionic and lactic, butyric and iso-butyric, valeric, and iso-valeric, caproic and iso-caproic, and heptanoic acids. VFAs are volatile and are subject to rapid microbial decomposition under aerobic conditions. Production of phytotoxic quantities of VFAs during composting (prior to compost maturation) are know to occur. The VFAs are most likely to be involved in odorous emissions when vegetative matter is present, such as occurs in the first stages of a composting operation when grass and green matter are delivered and sorted. They are unlikely to occur with biosolids alone.

Amines. These compounds can be produced in easily detectable quantities during high temperature processes. In composting, amines result from microbial decomposition that involves decarboxylation of amino acids. The amines that are produced are easily volatilized when temperatures are elevated above about 27°C. In biosolids produced with polymeric flocculating agents, high ambient temperatures can accentuate volatilization of amines that may be microbially split off from the core backbone of the polymer. Amines include: methylamine, ethylamine, trimethylamine, and diethylamine. Amines often accompany ammonia emissions, and if chlorine is used chloramines may be released.

Table A-1. Range of Odor Thresholds for Selected Sulfur Compounds, Ammonia, and Trimethylamine as reported in the literature †

COMPOUND	ODOR CHARACTER	A μg/l	B μg/l	C μg/l	D μg/l	E μg/l	F μg/l	G μg/l
Hydrogen Sulfide	Rotten eggs	0.47	0.47	4.70	0.5 - 10.0	4.8	0.50	8.1
Dimethyl Sulfide	Decayed cabbage	0.10	1.00	3.00	2.5 - 50.8	1.00	1.00	--
Dimethyl Disulfide	Vegetable sulfide	--	1.00	--	0.1 - 346.5	--	--	--
Methyl mercaptan	Sulfidy	1.10	1.10	0.50	4.0×10^5 -82	2.10	0.50	1.6
Ammonia	Pungent, irritating	--	37.0	470	26.6 -39,600	46,800	17,000	5,200
Trimethylamine	Fishy, pungent	--	--	--	0.8	0.21	--	0.2

† Letters correspond to the references cited as follows: A = Bowker et al. 1989; B=Versucheren, 1996; C=National Research Council, 1979; D=Ruth, 1986; E=Leonardos et al., 1969; F=Buonicore and Davis, 1992; G= Amoore and Hautala, 1983.

TableA-2. Selected odorous compounds observed in association with manure, compost, sewage sludge and biosolids as reported in the literature with corresponding ranges of odor threshold values †‡

Compound	Odor Character	Odor Threshold $\mu l/1$ ($\mu g/l$)
Nitrogenous compounds		
Ammonia	Sharp pungent	5.2 ‡ (150)
Butylamine	Sour, ammonia-like	1.8 ‡ (6200)
Dibutylamine	Fishy	(0.016)†
Diisopropylamine	Fishy	1.8 ‡ (1300)
Dimethylamine	Putrid, fishy	0.13 (470)
Ethylamine	Ammonical	0.95 ‡ (4300)
Methylamine	Putrid, fish	3.2 ‡ (2400)
Triethylamine	Ammonical, fishy	0.48‡ (0.42)
Trimethylamine	Ammonical, fishy	0.00044 ‡
Nitrogenous Heterocyclics		
Indole	Fecal, nauseating	(0.00012 - 0.0015)†
Skatole	Fecal, nauseating	(0.00035 - 0.0012)†
Pyridine	Disagreeable,burnt pungent	0.17‡ (0.95)
Sulfur-containing compounds		
Dimethyl sulfide	Decayed vegetables	(0.0003 - 0.016) †
Diphenyl sulfide	Unpleasant	(0.0026) †
Dimethyl disulfide	Vegetable sulfide	(1.00)†
Hydrogen sulfide	Rotten eggs	8.1‡ (0.000029)
Sulfur dioxide	Pungent, irritating	1.1‡ (0.11)
Amyl mercaptan	Unpleasant, putrid	(0.0003)†
Allyl mercaptan	Strong garlic, coffee	(0.000005)†
Benzyl mercaptan	Unpleasant, strong	(0.013)†
Crotyl mercaptan	Skunk-like	(0.00000043)†
Ethyl mercaptan	Decayed cabbage	0.00076‡ (0.0000075)
Methyl mercaptan	Decayed cabbage, sulfidy	0.0016 ‡ (0.000024)
Propyl mercaptan	Unpleasant	0.0000025 - 0.000075
n-butyl mercaptan	Skunk, unpleasant	0.00097 (0.000012)
Thiocresol	Skunk, rancid	(0.0001)†
Thiophenol	Putrid, garlic-like	(0.00014)†
Other chemicals or compounds		
m-Cresol	Tar-like, pungent	0.000049-0.0079 (37)
n-butyl alcohol	Alcohol	0.84‡
Chlorine	Pungent, suffocating	0.31‡ (0.0020)
Acetaldehyde	Pungent fruity	0.050 ‡ (0.034)

† O'Neill and Phillips, 1992; Vesilind et al., 1986; converted from weight by volume concentration (mg/m³) to $\mu g/l$
‡ Amoore and Hautala, 1983; $\mu l/l$ is the odor threshold for dilutions in odor-free air, and $\mu g/l$ is the odor threshhold; both units are equivalent to parts per million.

Odor Determination

Odor Sample Collection

The need for odor sample collection is most likely to occur in the case of a longer term, constructed storage facility that has been unable to resolve odor emissions. In such a case, the facility operators may seek a more analytical approach upon which to base a remediation program. The proper collection of an air sample containing odorous compounds is essential for accurate analysis of the source of the odor. This is true for both qualitative and quantitative methods of odor analysis. The composition of an odor can range from a single chemical compound to a complex mixture of compounds. The components of the odor will often dictate the method of sampling. Therefore, insight as to which compounds or type of compounds may be contributing to the odor is desirable. Without this, a sampling method that can handle a broad range of compounds would be necessary. After identifying the type or group of odorants present, an appropriate sampling method can be used.

Several aspects should be considered when choosing an appropriate sampling method. The physical and chemical properties of the odorant will often determine which sampling method is desirable. Some of these properties are the polarity, volatility, and stability of the chemical compounds associated with the odor. To analyze the sample accurately, the composition of the odorant(s) must remain intact during sample collection. Condensation, adsorption, or permeation of the odorous compounds through the walls of the collection system can cause errors. For example, the boiling point of DMDS is 109°C, which means it is a liquid at ambient temperature. This physical property greatly influences DMDS emissions and measurement: elevated temperatures will dramatically increase DMDS emissions. When measuring DMDS and other compounds with high boiling points, it is important not to use sampling techniques that allow the sample to cool before it enters the analytical detector. Otherwise, these compounds will condense on the interior of the sample container, such as tedlar bags, and results will be negatively biased.

There are two main types of sources that are the focus of air sampling, area sources (such as from a pile) and point sources (such as from a stack); point sources can be more reliably sampled than area sources. At a biosolids storage site and its surrounding neighborhoods, ambient (outdoor) air would typically be the source for sample collection. The odors may still be intense (strong) even though the odorants are less concentrated at increasing distances from the facility. If scrubbers are used, stack emission samples are collected in the stack after scrubbers.

Odor samples can be collected in canisters, Tedlar bags, flux chambers, and adsorbent tubes. Adsorption tubes filled with Tenax packing and/or activated carbon are the most common types of traps used for ambient air sampling. Industrial hygienists often utilize specific adsorbent tubes for on-site analysis of specific individual compounds like ammonia, hydrogen sulfide, etc.

Sample Analysis

The ability to detect, identify, and quantify odorants in biosolids and other stored materials is an essential tool in the study of odors and in the development of prevention and mitigation treatments. If there is some correlation between the concentration of odorants found by an analytical method and the odor itself, then this tool is most useful. Since some odorants have low odor thresholds, the detection limit of an analytical method must be low or the odorants must be concentrated prior to analysis. The odorants and their concentrations in a sample will influence the choice of a method of analysis. The sampling approaches described below cover the range of simple, rapid, field methods for easy practical use through to the very complex instrumentally dependent methods, requiring laboratory analysis.

Sensory Odor Analysis

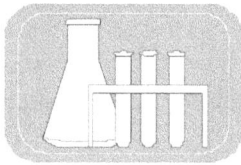

Characterizing the sensation experienced by inhaling an odorous sample is the object of a sensory odor measurement program. The human body experiences sensations, processes them, and then reacts. The olfactory system senses odor. Sensory analysis is most effective for samples containing complex mixtures of odorants or odorants at concentration levels below detection of an instrumental technique. It also produces simple, useful results that are meaningful to all concerned. Standardized testing protocols are now available for measuring odor intensity (ASTM E 544-75-88) and odor to threshold ratio (ASTM E679-91).

Odor Character Descriptors - In addition to the intensity of an odor, what an odor smells like is a big factor in determining whether it is objectionable. What an odor smells like is called the odor character and can be described through the use of various descriptors--words or phrases that most accurately represent the quality of the particular odor of concern. Each panelist is asked to describe the odor that was sensed. The problem with odor descriptors like "sweet," "musty," "sour," "putrid," "rotten," etc. is that different individuals may use a variety of words or phrases to describe the same odor. Even using what is called a "Hedonic Scale," which provides the panelist with a numbered scale or one with odor descriptors already provided, does not eliminate the human factor and the subjective nature of odor relative to its effect on different individuals.

Trained Odor Investigators - An extension of the use of simple odor descriptors is the odor patrol which utilizes trained odor investigators--people who have been trained to detect odor intensities. These people have "calibrated" their noses to certain odor intensities. They are trained to go "on site" and rate the odor intensity on a numeric scale. (see Chapter 2 for examples). Some examples of the types of written reports used for record keeping on-site and for citizen odor complaints appear at the end of this Appendix.

Scentometer - For direct field measurement of dilution-to-threshold, this hand-held device is sometimes used. Varying proportions of ambient (odorous) air, drawn through a activated carbon filter, are introduced to an individual"s nose. The ratio of ambient air to filtered air at which the individual detects an odor becomes the dilution-to-threshold. Odor inspectors using this method require training and experience so they can develop confidence in its application. This device has been used successfully by some inspectors in a few states.

Olfactometry - An olfactometer with an odor panel is another way to conduct a sensory analysis of odorous air samples. An olfactometer is an apparatus that presents an air sample containing the odorous component to an individual at varying dilutions with odor-free air. The object is to determine what level of dilution is necessary for each panelist to begin to detect an odor. From a series of these exposures, results for the odor panel can be calculated. These results can be expressed in the form of an odor to threshold ratio, or dilution level required for a percentage of the panel to detect the odor.

The Butanol Wheel - The intensity of an odor is also an important parameter when measuring odors. However, since the characteristic odors of various compounds are so different, it is difficult for individuals to compare the relative strengths or intensities of different odors. This can be overcome by using a reference compound to which the odor strengths can be compared. In this way, odors can be analyzed so that individuals not subjected to the actual odors can understand the results. The reference compound that is most widely used is n-butanol. A Butanol Wheel (2 - Procedure A) is used to measure the intensity (strength) of an odor by this comparative method.

The Butanol Wheel is similar to the olfactometer because it delivers the odorous compound and dilution air into ports to make different dilutions. The odorous compound in this case is the butanol vapor. The intensity of an odorous sample is measured by determining at what dilution level of the Butanol Wheel the sample matches the strength of the butanol vapor. An odor panel (group of people, each one exposed to the odor sample and butanol reference independently) is used to make the comparisons. By calculating the dilution of n-butanol vapor to which the odorous sample is equivalent, it is possible to express the intensity of the unknown odor in terms of a known intensity.

One of the principal differences between the forced-choice ascending concentration and the butanol wheel methods is that in the latter the odorous sample is tested at full strength against a series of diluted standards, whereas in the olfactometer method, the odorous sample itself is diluted as it is being evaluated. This difference results in assessment of odor intensity as well as dilution threshold ratio, two different sensory characteristics of the odor. This makes these two sensory test methods complementary to each other.

Chemical Analyzers and Instruments

There are many instruments and methods that can accurately measure odorous compound concentrations. One that combines sampling and analysis is a hand-held reactive absorbent tube, which is available for ammonia, hydrogen sulfide, and several other compounds of concern to industrial hygienists. There are single compound analyzers, such as a hydrogen sulfide (H_2S) meter, that measures one analyte. Multiple compound analyzers, like a gas chromatograph (GC), can measure more than one analyte. There are specific detectors for a GC that are sensitive to certain types of compounds. If these types of compounds are unknown or their mixture is complicated, then a mass spectrometer detector and an electronic library of compounds is necessary. The latter is an expensive and sophisticated analytical approach and one that is usually reserved for a research setting, not typically routine monitoring.

References

ASTM. 1968. Basic principles of sensory evaluation. ASTM Special Technical Publ., No. 433. Amer. Soc. For Testing and Materials. Philadelphia, PA.

ASTM. 1989. Standard recommended practices for referencing supra-threshold odor intensity. E544-75(88). Annual Book of Standards, Vol. 11.5. Amer. Soc. For Testing and Materials, Philadelphia, Pa.

ASTM. 1991. Standard practice for determination of odor and taste thresholds by a forced-choice ascending concentration series methods of limits. E679-91. 1991. Annual Book of Standards, Vol. 11.5. Amer. Soc. For Testing and Materials. 5 p.

Barnebey & Sutcliffe Corporation. 1974. Scentometer: An Instrument for Field Odor Measurement. Columbus, OH

Borgatti, D., G.A. Romano, T.J. Rabbitt, and T.J. Acquaro. 1997. The 1996 Odor Control Program for the Springfield Regional WWTP. New England WEA Annual Conf., 26-29 January 1997, Boston, MA.

Bowker, R.P.G., J.M. Smith, and N.A. Webster. 1989. Odor and corrosion control in sanitary sewerage systems and treatment plants. Noyes Data Corp., Park Ridge, N.J., U.S.A.

Bruvold, W.H., S.M. Rappaport,T.C. Wu, B.E. Bulmer, C.E. DeGrange, and J.M. Kooler. 1983. Determination of nuisance odor in a community. J. Water Pollut. Control Fed. 53:229-233.

Bruvold, W.H. Laboratory panel estimation of consumer assessments of taste and flavor. J. Appl. Psychol. 54: 326

Buonicore, A.J. and W.T. Davis (eds.). 1992. *Air pollution engineering manual* . Air & Waste Management Association. Van Nostrand Reinhold, NY.

Dravnieks, A. 1985. Atlas of odor Character Profiles, sponsored by Section E-18.04.12 on Odor Profiling of Subcommittee E-18.04 on Instrumental-Sensory

Relationships, ASTM Committee E-18 on Sensory Evaluation of Materials and Products. Philadelphia, PA.

Hentz, L. H. 1997. The Chemical, Biological and Physical Origins of Biosolids Emissions: A Review, Post, Buckley, Schuh & Jernigan, Inc. Bowie, MD.

Leonardos, G., D. Kendall, and N. Barnard. 1969. Odor Threshold determinations of 53 odorant chemicals. Air Pollut. Control Assoc. J. 19(2):91-95.

Lue-Hing, C., D.R. Zenz, and R. Kuchenrither. *1992. Municipal Sludge Management - Processing, Utilization and Disposal*, Water Quality Management Library (Volume 4), Technomic Pub Co., Inc. Lancaster, PA.

Miedema, H.M.E. and J.M. Ham. 1988. Odour annoyance in residential areas. Atmos. Environ. 2:2501-2507.

National Research Council. 1979. *Odors from Stationary and Mobile Sources.* National Acad. Sci., Washington, D.C.

Rosenfeld, P. 1999. Characterization, Quantification, and Control of Odor Emissions from Biosolids Application to Forest Soil. Ph.D. Dissertation. University of Washington, Seattle, WA.

Ruth, J.H. 1986. Odor thresholds and irritation levels of several chemical substances: A Review. Am. Ind. Hyg. Assoc. J. 47:A142-A151.

U.S. EPA. 1973. National Survey of the Odor Problem, Phase III. A Study of the Social and Economic Impact of Odors. La Jolla California, Copley Intl. Corp., EPA Report No. EPA-650/5-73-001, EPA, RTP. Phase I, 1970 , Phase II, 1971.

Verschueren, K. 1996. Handbook of environmental data on organic chemicals . 3rd ed. Van Nostrand Reinhold, NY. 2064 p.

Vesilind, P. A., Hartman, G. C., and Skene, E.T. 1986. *Sludge Management and Disposal for the Practicing Engineer*, Lewis Publishers, Inc., Chelsea, MI

Wilby, F.V. 1969. Variation in recognition odor threshold of a panel. J. Air Pollut. Contr. Assoc. 19(2):96-100.

Winneke, G. and J. Kastka. 1977. Odor pollution and odor annoyance reactions in industrial areas of the Rhine-Ruhr region, pp. 471-479. *In* Le Magnead MacLeod. (Ed.), *Olfaction and Taste*. IV. London

Yonkers. 1997. Process compatibility testing D. Odor. In Specifications for Furnishing and Delivering Liquid Emulsion type polymer (40-50 percent active) for Centrifuge dewatering of sludge. Yonkers Joint WWTP, Ludlow Dock, South Yonkers, NY.

SPRINGFIELD, MA ODOR NOTIFICATION FORM

The purpose of this form is to identify odors than can potentially migrate off Bondi Island, where the Springfield WWTP is located and to communicate those observations to the respective island facilities. Such a form could be applied to a large field storage site.

NOTIFIER/PHONE_____/_____

Odor Date/Time _____ Strength: weak, moderate, strong

Location of Odor _____

Temperature:_____ Wind speed/direction_____

Source	Odor Type Detected			
WWTP	Primary Treatment	Secondary Treatment	Biosolids	Other
Incinerator	Smoke	Ash	Hopper Juice	Other
Cover Tech	Leaf/Earthy	Yard Waste	Raw Paper Sludge	Other
Landfill Gas	Natural Gas	Other		
RCI Landfill	Sludge	Other		
RCCI Compost	Compost	Other		
Waste Stream	Sludge	Ammonia		
Street BioFilter	Chemical	Sewage		

Odor Descriptors: (check all that apply) ☐ sewer ☐ putrid foul decayed ☐
chemical fecal (like manure) ☐ garbage truck ☐ rotten eggs ☐ burnt ☐ smoky
☐ musty earthy

Source contacted _____; Source copied _____

Message left _____;Senior Operator_____

Odor confirmed by Sr. Operator?   Yes   No

Comments:

_____

_____

_____

_____

_____

_____

_____

_____

RESIDENT ODOR COMPLAINT FORM
Courtesy of Springfield Regional WWT Facility

Date / Time of Odor _____ AM   PM

Wind Direction / Speed _____

Air Temperature / Relative Humidity _____

Weather Conditions _____

Senior Operator _____

RESIDENT INFORMATION
Name _____

Address _____

       City _____

Zip Code _____ Telephone No. _____

Odor Description (circle all applicable) sewer   putrid foul decayed   chemical
fecal (like manure)   garbage truck   rotten eggs   burnt smoky   musty earthy

Duration / Frequency of Odor
_____

Intensity of Odor   Weak   / Moderate / Strong

Senior Operation Information (Detailed)

_____

_____

_____

_____

_____

_____

_____

EXAMPLE OF PERFORMANCE STANDARDS FOR ODOROUS EMISSIONS FROM A PERMANENT CONSTRUCTED FACILITY

Example adapted from a Compost Site Conditional Use Permit, courtesy of St. Croix Sensory, Inc.

Odor Testing

1. This odor testing practice references the odor intensity of the ambient air to an "Odor Intensity Referencing Scale (OIRS)".

2. The odor of the ambient air is matched (ignoring differences in odor quality) against the OIRS (see Section B in the following section) by trained inspectors. The inspector reports that point, or in between points, on the reference scale which, in her(his) opinion, matches the odor intensity of the ambient air.

3. The procedure followed for field odor testing is in accordance with Procedure B - Static-Scale Method of ASTM E-544, except for the following adaptations:

   a.   The geometric progression scale ratio = 3.

   b.   Use screw-cap containers for reference concentrations of butanol in water.

   c.   Inspectors may memorize the OIRS.

   d.   Inspectors may use a charcoal filter, breathing mask to avoid olfactory adaption (fatigue) in the ambient air.

   e.   Inspectors sniff ambient air and match its intensity to the reference scale.

   f.   Inspectors breathe charcoal filtered air for three minutes in between snifffings of ambient air.

   g.   Odorous air sampling shall be performed upon the complainant's property. The inspector shall not be accompanied by the complainant and results shall be released after a written report is filed. The inspector shall not conduct the odorous air sampling if the complainant is present.

   h.   The inspector shall also sample the ambient air immediately upwind from the compost site to determine the presence and level of any odors entering the site from other sources. These records and observations shall be a part of the written report

   I.   The Odor Intensity Referencing Scale (OIRS) will use numbers and descriptions corresponding to butanol concentrations as indicated below:

No.	Category Description	N-Butanol (ppm) In air/ in water
0	No Odor	0/ 0
1	Very Faint	25/ 250
2	Faint	75/ 750
3	Distinct, Noticeable	225/ 2250
4	Strong	675/ 6750
5	Very Strong	2025/ 20250

Reasonable operating conditions will allow for X (a designated number) or fewer recorded sniffings by an inspector of the ambient air over a period of Y minutes with a geometric average OIRS value of:

a) 3.0 or greater if there is a permanent residence upon the property, or,
b) 4.0 or greater if the property does not contain a permanent residence.

Appendix B

Pathogens

Transmission of Pathogens

Pathogen levels in wastewater reflect the presence or absence and level of pathogens in the general population served by the municipal facility. Wastewater treatment processes are designed to reduce the presence of pathogens in treated discharge water. In addition, there are several treatment processes that are used to reduce the pathogen content in the residual solids. The Part 503 rules specify these pathogen limits for two classes of pathogen reduction, Class A and Class B, in treated solids (see Tables B-2 and B-3).

In assessing the disease potential of biosolids or of a storage situation, the amounts of pathogen present as well as the potential routes of infection, the likelihood of a person contacting the source of the pathogen, the success of storage containment, as well as the amount that a person would potentially ingest or inhale if containment failed, and the virulence of the disease agent must all be taken into account. This type of information is essentially the same as that used to assess the disease potential of infectious pathogens that we contact in our daily activities (involving hand-to-hand, hand-to-eye, hand-to-mouth contact with pathogen sources, or inhalation and/or ingestion). It is clear from our knowledge of daily activity exposures that only some exposures result in disease. This may in part be attributed to the fact that some are more intense than others, e.g., the intense exposure to air in enclosed areas like commercial aircraft cabins, movie theatres, schools, and daycare centers; or, food and beverages prepared, imported, and/or served commercially by persons carrying and possibly transmitting a variety of microorganisms; or simply hand shakes with friends and colleagues. When disease occurs, we know that the amount of the pathogens present, their virulence, the person's susceptibility, and the exposure route were all sufficiently above the threshold levels that result in an infection. Fortunately, most daily activities do not result in disease.

During the course of wastewater treatment, the microorganisms in sewage are reduced in number, and become concentrated in the solids. Untreated (unstabilized) solids have a greater potential to contain significantly larger amounts of pathogens than do solids that have been treated with pathogen reduction processes that result in Class A or Class B biosolids according to Part 503 rules. Class A biosolids have no detectable pathogens, whereas Class B biosolids have significantly reduced levels of pathogens. Hence, the part 503 rule specifies site access and crop harvesting restrictions for Class B biosolids so they can be safely land applied. For these reasons, it is recommended that only Class A or B biosolids intended for land application be brought to field sites/facilities for storage.

Table B-1. Major Pathogens Potentially Present in Municipal Wastewater and Manure*

Bacteria	Disease/Symptoms for Organism
Salmonella spp.	Salmonellosis (food poisoning), typhoid
Shigella spp.	Bacillary dysentery
Yersinia spp.	Acute gastroenteritis (diarrhea, abdominal pain)
Vibrio cholerae	Cholera
Campylobacter jejuni	Gastroenteritis
Escherichia coli (enteropathogenic)	Gastroenteritis
Viruses	
Poliovirus	Poliomyelitis
Coxsackievirus	Meningitis, pneumonia, hepatitis, fever, etc.
Echovirus	Meningitis, paralysis, encephalitis, fever, etc.
Hepatitis A virus	infectious hepatitis
Rotavirus	Acute gastroenteritis with severe diarrhea
Norwalk Agents	Epidemic gastroenteritis with severe diarrhea
Reovirus	Respiratory infections, gastroenteritis
Protozoa	
Cryptosporidium	Gastroenteritis
Entamoeba histolytica	Acute enteritis
Giardia lamblia	Giardiasis (diarrhea & abdominal cramps)
Balantidium coli	Diarrhea and dysentery
Toxoplasma gondii	Toxoplasmosis
Helminth Worms	
Ascaris lumbricoides	Digestive disturbances, abdominal pain.
Ascaris suum	Can have symptoms: coughing, chest pain.
Trichuris trichiura	Abdom. pain, diarrhea, anemia, weight loss
Toxocara canis	Fever, abdominal discomfort & muscle aches
Taenia saginata	Nervousness, insomnia, anorexia.
Taenia solium	Nervousness, insomnia, anorexia.
Necator americanus	Hookworm disease
Hymenolepis nana	Taeniasis

* Not all pathogens are necessarily present in all biosolids and manures, all the time.

Methods for Meeting 40 CFR 503 Pathogen Requirements

The U.S. EPA 40 CFR 503 regulations, specifically 503.32(a) and (b), require biosolids intended for agricultural use to meet certain pathogen and vector attraction reduction conditions. The intent of a Class A pathogen requirement is to reduce the level of pathogenic organisms in the biosolids to *below detectable levels.* The intent of the Class B requirements is to ensure that pathogens have been reduced to levels that are unlikely to pose a threat to public health and the environment under the specific use conditions. For Class B material that is land applied, site restrictions are imposed to minimize the potential for human and animal contact with the biosolids for a period of time following land application until environmental factors have further reduced pathogens. No site restrictions are required with Class A biosolids. Class B biosolids cannot be sold or given away in bags or other containers. The criteria for meeting Class A requirements are shown in Table B-2, and criteria for Class B are shown in Table B-3.

Table B-2. Criteria for Meeting Class A Requirements

Parameter	Unit	Limit
Fecal Coliform or Salmonella	MPN/g TS*	1000
	MPN/4g TS	3
AND, one of the following process options		
Temp/Time based on % Solids	Alkaline Treatment	
Prior test for Enteric Virus/Viable Helminth	Post test for Enterec Virus/Viable Helminth Ova	
Composting	Heat Drying	
Heat Treatment	Thermophilic Aerobic Digestion	
Beta Ray Irradiation	Gamma Ray Irradiation	
Pasteurization	PFRP** Equivalent Process	

* Most probable number per gram dry weight of total solids
** Process to Further Reduce Pathogens; see Glossary in this document, and the EPA, Plain English Guide to Part 503

Table B-3. Criteria for Meeting Class B Requirements.

Parameter	Unit	Limit
Fecal Coliform	MPN or CFU/g TS*	2,000,000
OR, **one of the following process options**		
Aerobic Digestion	Air Drying	
Anaerobic Digestion	Composting	
Lime Stabilization	PSRP** Equivalent	

* Most probable number or colony -forming units per gram dry weight of total solids
** Process to Significantly Reduce Pathogens; see Glossary in this document, and the EPA, Plain English Guide to Part 503

Vector Attraction Reduction (VAR)

Under Subpart D of the Part 503 rule, safety and health protection with regard to biosolids management requires that biosolids meet one of 12 options to demonstrate vector attraction reduction, VAR (specifically 503.33). Options 1 - 8 consist of operating conditions or test to demonstrate VAR in treated biosolids, whereas options 9 - 11 use the soil as a barrier to prevent vectors from coming in contact with the biosolids. Materials that meet VAR 1 - 8 at the WWTP require less management at the storage site than biosolids without VAR treatment. All Class B biosolids that are stored require the same level of protection by site management as are provided by Class B site restrictions for land application.

Options prescribed for VAR are shown in Table B-4, and although these are not federally binding on biosolids storage operations, they do apply to the biosolids that are released from storage for land application. Some of these options rely on reducing the volatile solids in biosolids, and this can contribute to increased stability of the material, which is often associated with odor reduction. Furthermore, proper storage can assist in volatile solids reduction and as such in meeting vector attraction reduction requirements applicable to the use and disposal of biosolids according to Part 503.

The descriptions of the VAR methods presented in the regulation are treatment standards and descriptions only, but additional guidance is available (see EPA, 1992, EPA, 1995, Farrell et. al. 1996 in Chapter 4 references) which explains the rationale for the options. Also, Smith et. al. (1994) in another EPA guidance document provide direction on sampling and testing protocols.

Table B-4. Summary of Requirements for Vector Attraction Reduction Options.

Option	Requirement	Where/When Requirements must be met
1 Volatile Solids (VS) Reduction	$\geq$ 38% VS reduction during solids treatment	Across the process
2 Anaerobic benchscale test	< 17% VS loss, 40 days at 30°C to 37°C (86°F to 99°F)	On anaerobic digested biosolids
3 Aerobic benchscale test	< 15% VS reduction, 30 days at 20°C (68°F)	On aerobic digested biosolids
4 Specific Oxygen Uptake Rate	SOUR at 20°C (68°F) is $\leq$ 1.5 mg oxygen/hr/g total solids	On aerobic stabilized biosolids
5 Aerobic Process	$\geq$14 days at > 40°C (104°F) with an average > 45°C (113°F)	On composted biosolids
6 pH adjustment	$\geq$ 12 measured at 25°C (77°F)*, and remain at pH > 12 for 2 hours and $\geq$11.5 for 22 more hours	When produced or bagged
7 Drying without primary solids	$\geq$ 75% Total Solids (TS) prior to mixing	When produced or bagged
8 Drying with primary solids	$\geq$ 90% TS prior to mixing	When produced or bagged
9 Soil Injection	No significant amount of solids is present on the land surface 1 hour after injection. Class A biosolids must be injected within 8 hours after the pathogen reduction process.	When applied
10 Soil Incorporation	$\leq$ 6 hours after land application; Class A biosolids must be applied on the land within 8 hours after being discharge from the treatment process.	After application
11 Daily cover at field site	Biosolids placed on a surface disposal site must be covered with soil or other material at the end of each operating day.	After placement
12 pH adjustment of septage	$\geq$ 12 measured at 25°C (77°F)*, and remain at $\geq$ 12 for 30 minutes without addition of more alkaline material.	Septage

* or corrected to 25°C

References

see Chapter 4 References.

Appendix C

Runoff Management Practices

Water Management

Field stockpiles and constructed storage facilities with vehicle access points at or below ground-level such as concrete pads may be vulnerable to surface runoff. Methods for managing surface runoff include:

- Careful site selection and evaluation to assess expected volume of precipitation and runoff during planned storage periods, and optimizing topographic location to minimize exposure to runoff or flooding. Good site selection can frequently eliminate the need for additional runoff controls.

- Minimizing the amount of direct precipitation and upslope runoff encountering stored material through use of stormwater diversions, shaping of stockpiles roofing, or enclosing the facility.

- For constructed facilities, properly managing water that comes into contact with the residual material through collection of accumulated water, or for field stockpiles, use of filter strips and buffer zones.

- For constructed facilities, sumps or gravity flow can be used for transport of accumulated water to on-site filter strips or treatment ponds. Water can also be mixed with the residual for land application, decanted, and transported to off-site treatment facilities or irrigation systems (taking care not to exceed hydraulic loading rates to prevent ponding or run-off).

Best Management Practices

Grassed waterways: Are shaped and graded channels that are protected with vegetation, stone or other materials to carry surface water at a non-erosive velocity to a stable outlet. The vegetation in the waterway protects the soil from erosion caused by concentrated flows, while carrying water down slope. Grassed waterways may be used as outlets for diversions or to convey water to treatment ponds or filter areas. Waterways should be inspected periodically, any eroding areas should be repaired and they should be mown, reseeded and fertilized as needed to maintain good vegetative cover.

Provide stabilized machinery crossings, where needed, to prevent rutting of the waterway. Waterways should not be crossed when wet.

Silt Fence

Temporary barriers of woven geotextile fabric (approximately two feet high) are used to filter surface runoff, reduce its velocity and trap sediment from disturbed areas. Silt fences can only be used to intercept sheet flow, they cannot be used in swales or other areas where the flow of water is concentrated.

Silt fences are installed on or parallel to contours. To work effectively, the bottom of the entire length of the fabric must be placed in a trench or slot in the soil and back filled. This ensures a continuous seal with the ground, so that water and sediment will be trapped and not pass under the fence. To ensure that a silt fence is not knocked down or overwhelmed with sediment, the maximum length of a silt fence is proportional to slope steepness and length. Consult your local natural resource conservationist for specifications applicable to your site. Silt fence should be inspected after each rainfall event and maintained when bulges occur or when sediment accumulates to 50 percent of the fence height (See also Figure C-1).

Straw Bale Dikes

Straw bale dikes are temporary measures used to filter sediment from sheet flow runoff so that deposition of transported sediment can occur. Straw bale dikes clog and deteriorate rapidly and require frequent maintenance. Bales should be placed in a row on the contour with the ends of each bale tightly abutting the adjacent bales and securely anchored in place with stakes. Bales should be entrenched several inches in the soil to ensure a good seal with the ground to prevent water and sediment from flowing under the bales instead of through them (See also Figure C-2).

Filter Strips

A strip or area of grass or other vegetation that removes sediment, organic matter, nutrients and other pollutants from runoff and wastewater by filtration, infiltration, absorption, adsorption, decomposition and volatilization. In many cases there may be enough natural vegetation present to filter pollutants. If not, a filter area can be planted alone or in combination with existing natural vegetation. This practice may be applied downslope of long term stockpiles, or storage facilities, at the lower edges of fields or adjacent to streams, channels, or ponds.

Filter strips are designed to handle sheet flow of surface runoff. If any storm water management practice, such as a grassed waterway deliver water to these areas they must be designed with outlets that distribute and slow the concentrated flow of water into an even sheet across the top edge of the filter. Grassed filter strips are placed along the contour. They must be long enough and wide enough so that peak sheet flow does not exceed the maximum permissible depth (e.g. one-half-inch) and so that the time it takes the water to pass through the filter provides the necessary level of pollutant removal and treatment. Filter strips should be protected from damage by farm equipment and vehicle traffic. They should be inspected regularly to ensure that the area remains properly vegetated and that no gullies or areas of concentrated flow develop to short circuit the system. Any necessary reseeding or reshaping should be done promptly.

Berms/Earth Dikes

A temporary earthen ridge of soil, shaped along the contour and compacted, to divert runoff around a stockpile or constructed storage area. Berms intercept up-slope sheet flow and outlet to an undisturbed stabilized area or watercourse at a non-erosive velocity. For temporary stockpiles berms may be created with on-farm tillage equipment. Berms should be sized to the upslope drainage area. If necessary, depending on soil type and the expected length of storage, the berm should be stabilized by seeding or mulching. Berms should be regularly inspected and maintained to ensure they are not breached or eroded. Following removal of the field stockpile, berms should be removed and the area returned to its original grade (See also Figures C-3 and C-4).

Diversions

A channel constructed across a slope with a supporting ridge on the lower side used to divert clean runoff water away from a storage area. Diversions prevent clean runoff from coming into contact with stored biosolids and protect down-slope areas from erosion. A diversion must discharge runoff wafer to a stable outlet at non-erosive velocities. The outlet may be a grassed waterway, a vegetated area, or a stable watercourse. Diversions should be compacted and stabilized by seeding, and regularly inspected. Repair and reseed any bare areas immediately, keep channel and outlet clear of debris, keep burrowing animals out of the bank; mow, reseed, and fertilize as needed to maintain vegetation.

Heavy Use Protection

For long term stockpiles or permanent storage facilities, protect loading and other areas from erosion with gravel or paving, as necessary.

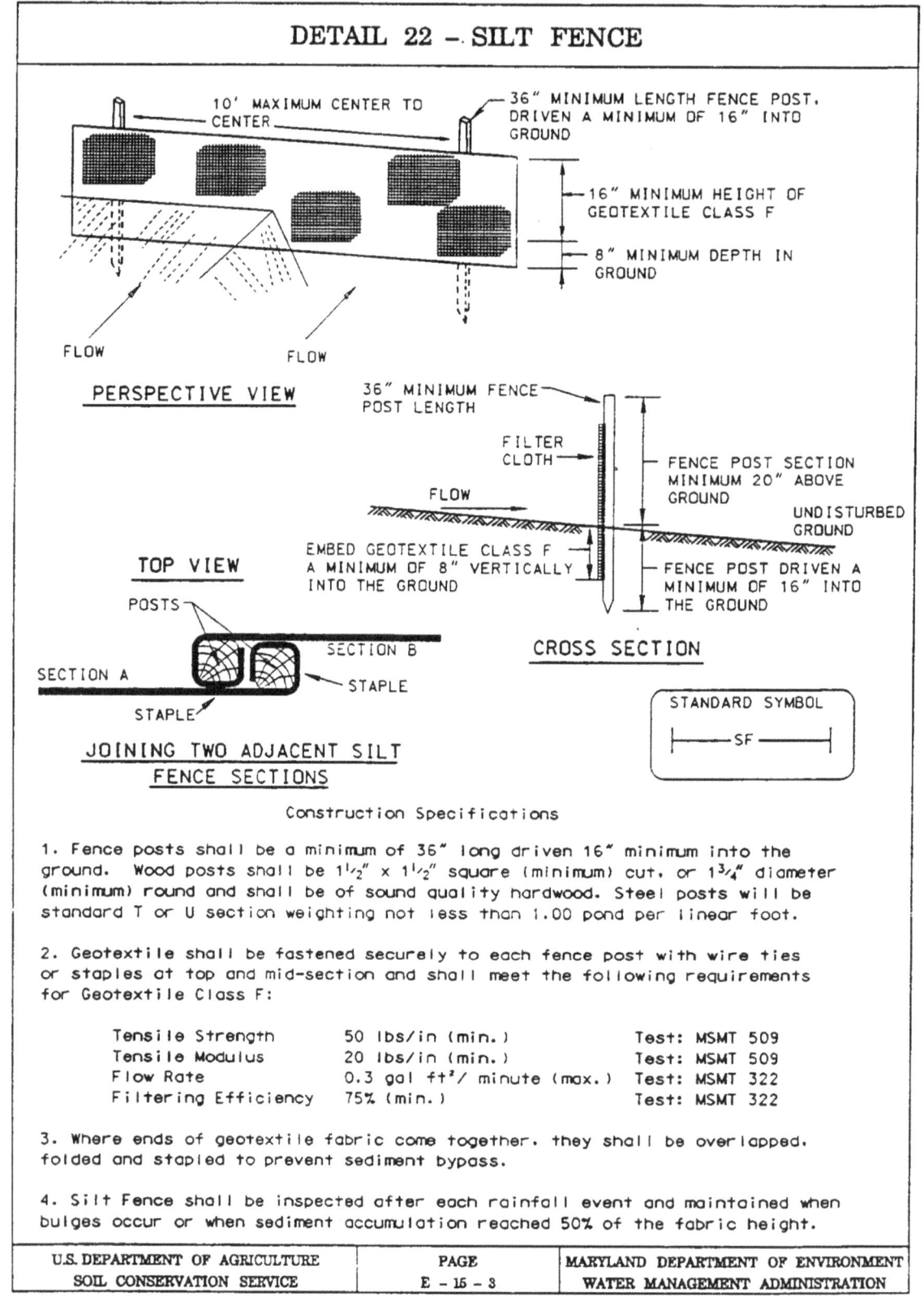

Figure C-1. Silt Fence Design Diagram

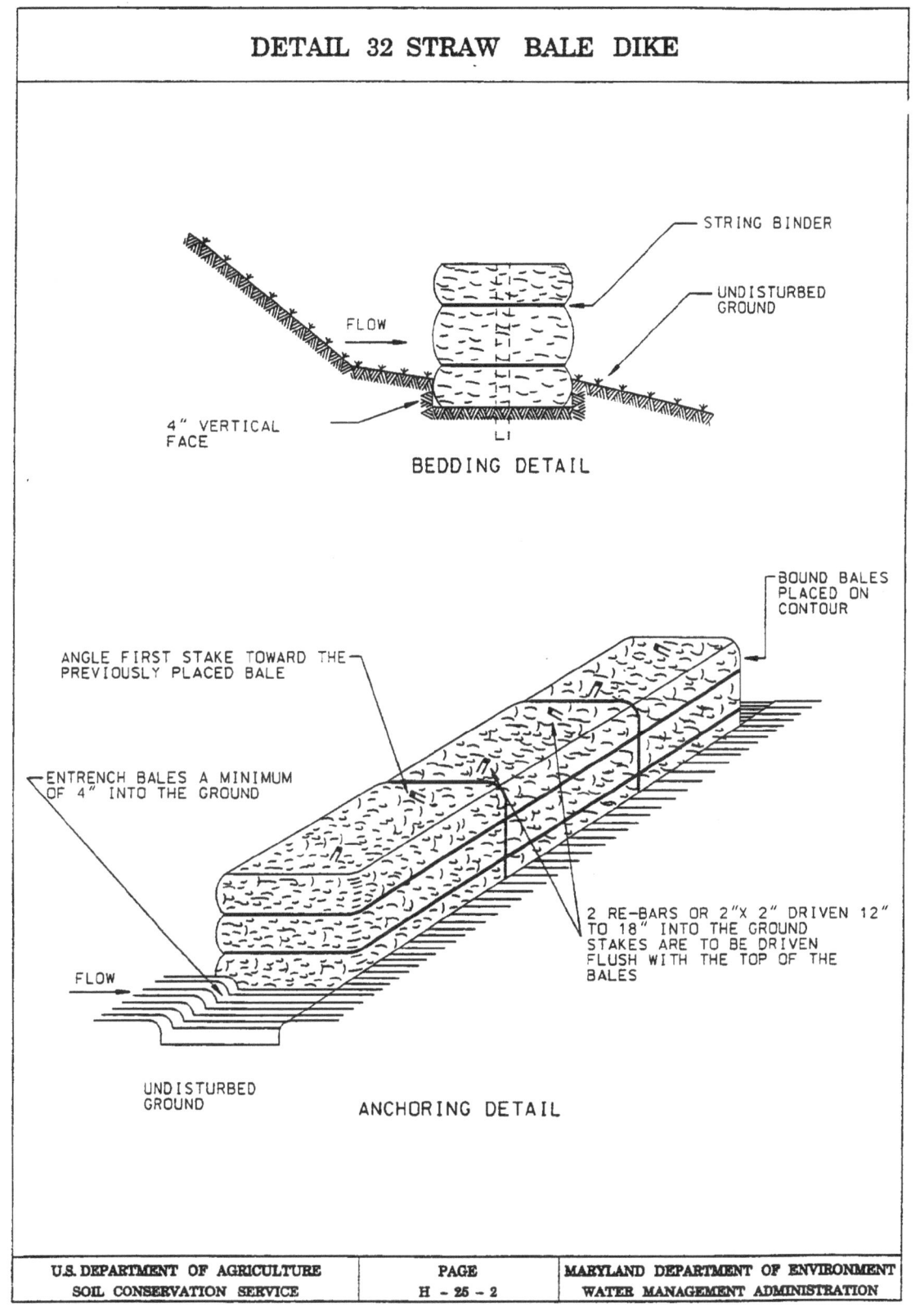

DETAIL 32 STRAW BALE DIKE

STRING BINDER

UNDISTURBED GROUND

FLOW

4" VERTICAL FACE

L¹

BEDDING DETAIL

BOUND BALES PLACED ON CONTOUR

ANGLE FIRST STAKE TOWARD THE PREVIOUSLY PLACED BALE

ENTRENCH BALES A MINIMUM OF 4" INTO THE GROUND

FLOW

UNDISTURBED GROUND

2 RE-BARS OR 2"X 2" DRIVEN 12" TO 18" INTO THE GROUND STAKES ARE TO BE DRIVEN FLUSH WITH THE TOP OF THE BALES

ANCHORING DETAIL

| U.S. DEPARTMENT OF AGRICULTURE SOIL CONSERVATION SERVICE | PAGE H – 25 – 2 | MARYLAND DEPARTMENT OF ENVIRONMENT WATER MANAGEMENT ADMINISTRATION |

Figure C-2. Straw Bale Dike Design Diagram

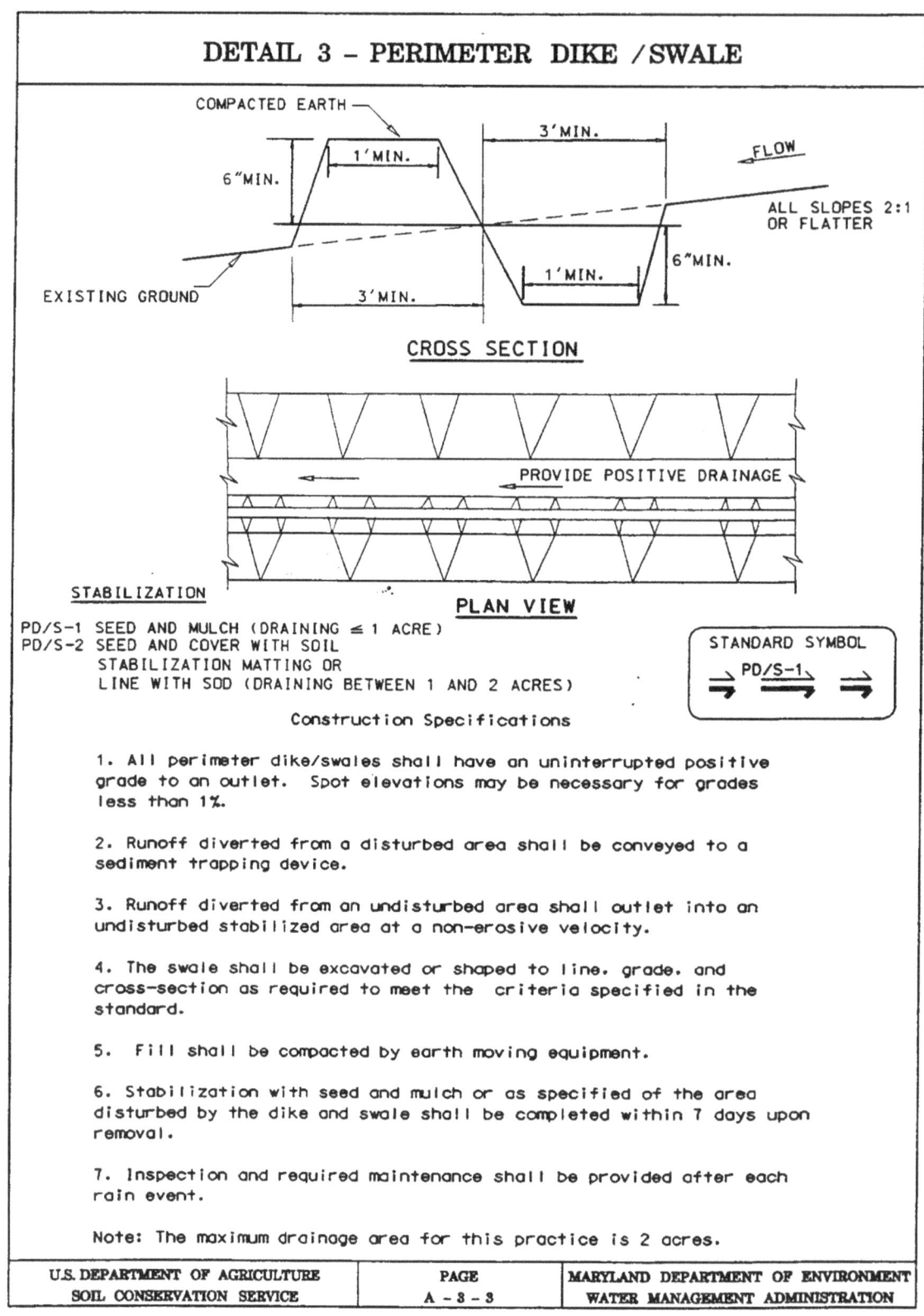

DETAIL 3 – PERIMETER DIKE /SWALE

CROSS SECTION

PLAN VIEW

STABILIZATION

PD/S-1 SEED AND MULCH (DRAINING ≤ 1 ACRE)
PD/S-2 SEED AND COVER WITH SOIL
        STABILIZATION MATTING OR
        LINE WITH SOD (DRAINING BETWEEN 1 AND 2 ACRES)

STANDARD SYMBOL
PD/S-1

Construction Specifications

1. All perimeter dike/swales shall have an uninterrupted positive grade to an outlet. Spot elevations may be necessary for grades less than 1%.

2. Runoff diverted from a disturbed area shall be conveyed to a sediment trapping device.

3. Runoff diverted from an undisturbed area shall outlet into an undisturbed stabilized area at a non-erosive velocity.

4. The swale shall be excavated or shaped to line, grade, and cross-section as required to meet the criteria specified in the standard.

5. Fill shall be compacted by earth moving equipment.

6. Stabilization with seed and mulch or as specified of the area disturbed by the dike and swale shall be completed within 7 days upon removal.

7. Inspection and required maintenance shall be provided after each rain event.

Note: The maximum drainage area for this practice is 2 acres.

U.S. DEPARTMENT OF AGRICULTURE SOIL CONSERVATION SERVICE	PAGE A – 3 – 3	MARYLAND DEPARTMENT OF ENVIRONMENT WATER MANAGEMENT ADMINISTRATION

1994

Figure C-3. Perimeter Dike/Swale Design Diagram

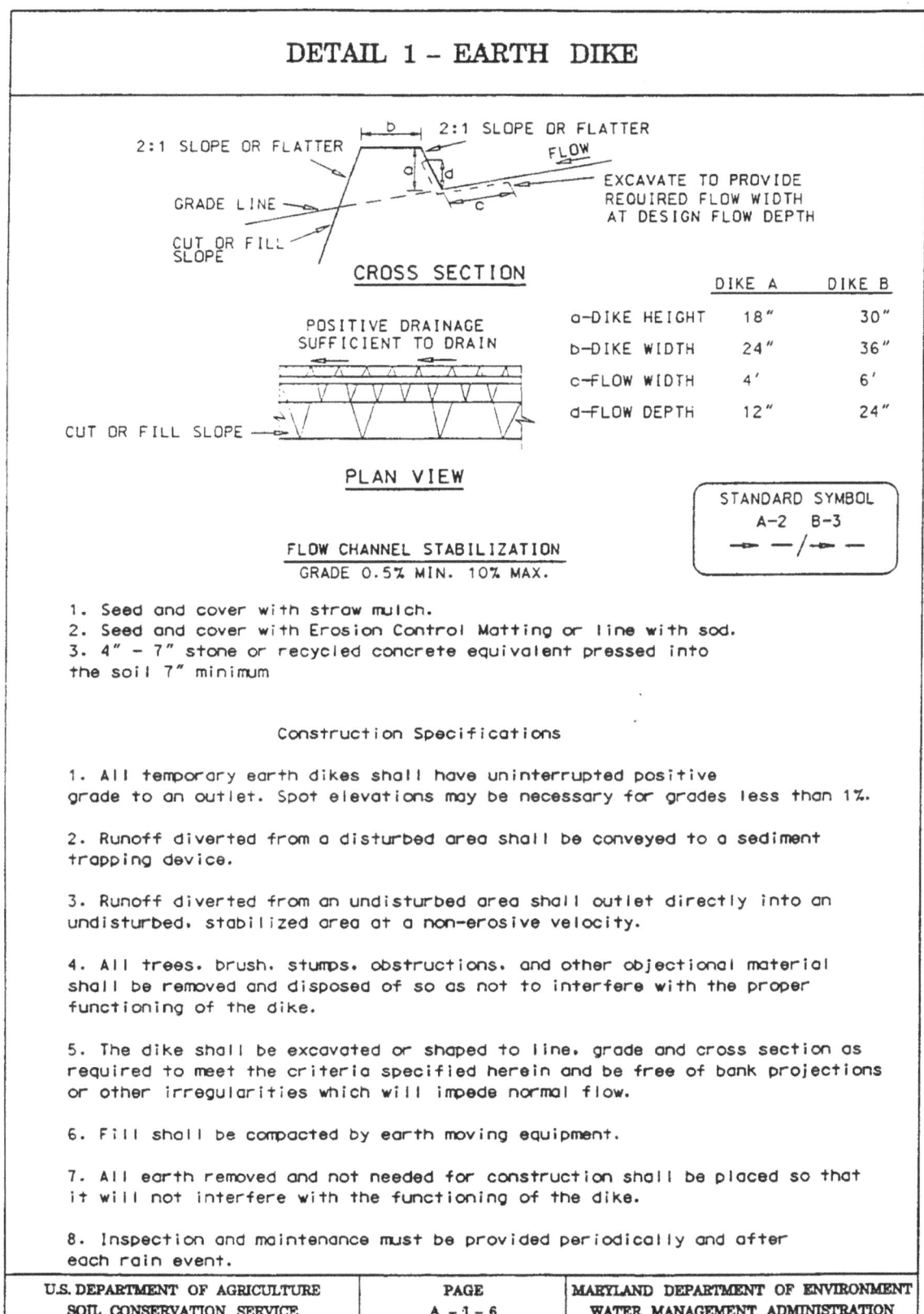

Figure C-4. Earthen Dike Design Diagram.

Natural Resources Conservation Service Regional Conservationists

Region	Name & Address
Eastern Office	Phone: 301-586-1387 or 1388 Calverton Office Bldg. #2 Suite 100 11710 Calverton Blvd. Beltsville, MD 20705
Midwest Office	Phone: 608-224-3010 2820 Walton Commons West Suite 123 Madison, WI 53704-6785
Northern Plains Office	Phone: 402-437-4082 100 Centennial Mall North Room 152, Federal Building Lincoln, NE 68508-3866
Southeastern Office	Phone: 404-347-6105 1720 Peachtree Rd., N.W. Suite 446N Atlanta, GA 30309-2439
South Central Office	Phone: 817-334-5224 501 W. Felix St., Bldg. 23 Felix & Hemphill Street Ft. Worth, TX 76115
Western Office	Phone: 916-491-2000 650 Capitol Mall, Room 7010 Sacramento, CA 95814

Appendix D

Nutrient Content of Organic By-Products

Proper soil and crop management is required to avoid contaminating surface or groundwater when using fertilizer materials. Plant nutrient requirements can be met by applying inorganic or organic fertilizers. Nutrient and carbon content information is also very useful when tailor blending products for specialty purposes.

Table D-1. Nutrient Content of Various Organic Materials*

Material	Percentage by Weight						
	N	P2O5	K2O	Ca	Mg	S	Cl
Apple pomace	2	—	0.2	—	—	—	—
Blood (dried)	12 - 15	3.0	—	0.3	—	—	0.6
Bone meal (raw)	3.5	22.0	—	22.0	0.6	0.2	0.2
Bone meal (steamed)	2.0	28.0	0.2	23.0	0.3	0.1	—
Brewers grains (wet)	0.9	0.5	—	—	—	—	—
Common crab waste	2.0	3.6	0.2	—	—	—	—
Compost (garden)	varies with feedstocks and processes						
Cotton waste from factory	1.3	0.4	0.4	—	—	—	—
Cottonseed meal	6 - 7	2.5	1.5	0.4	0.9	0.2	—
Cotton motes	2.0	0.5	3.0	4.0	0.7	0.6	—
Cowpea forage	0.4	0.1	0.4	—	—	—	—
Dog manure	2.0	10.0	0.3	—	—	—	—
Eggs	2.1	0.4	0.2	—	—	—	—
Egg shells	1.2	0.4	0.2	—	—	—	—
Feathers	15.0	—	—	—	—	—	—
Fermentation sludges	3.5	0.5	0.1	7.3	0.1	—	—
Fish scrap (dried)	9.5	6.0	—	6.1	0.3	0.2	1.5
Fly ash:							
coal	0.3	0.1	—	0.6	0.1	10.0	0.5
wood	9.8	—	0.7	—	—	—	—

Table D-1. Nutrient Content of Various Organic Materials (continued)*

Material	Percentage by Weight						
	N	P2O5	K2O	Ca	Mg	S	Cl
Frittercake:							
enzyme production	—	—	2.2	2.0	0.5	0.3	—
citric acid production	—	—	5.2	—	—	—	—
Garbage tankage	2.5	1.5	1.0	3.2	0.3	0.4	1.3
Greensand	—	1 - 2	5.0	—	—	—	—
Hair	12 - 16	—	—	—	—	—	—
Legume	3.0	1.5	1.0	0.5	2.4	1.9	1.2
Grass	0.8	0.2	0.2	0.3	0.2	—	—
Oak leaves	0.8	0.4	0.2	—	—	—	—
Oyster shell siftings	0.4	10.4	0.1	—	—	—	—
Peanut hull meal	1.2	0.5	0.8	—	—	—	—
Peat/muck	2.7	—	—	0.7	0.3	1.0	0.1
Pine needles	0.5	0.1	—	—	—	—	—
DAF sludge	8.0	1.8	0.3	—	—	—	—
Potato tubers	0.4	0.2	0.5	—	—	—	—
Potato, leaves & stalks	—	0.6	0.2	0.4	—	—	—
Potato skins, raw ash	—	—	5.2	2.0	7.5	—	—
Sawdust	0.2	—	0.2	—	—	—	—
Sea marsh hay	1.1	0.2	0.8	—	—	—	—
Seaweed (dried)	0.7	0.8	5.0	—	—	—	—
Sewage sludge (municipal)	2.6	3.7	0.2	1.3	0.2	—	—
Shrimp waste	2.9	10.0	—	—	—	—	—
Soot from chimney	—	0.5 - 11	—	1.0	0.4	—	—
Soybean meal	7.0	1.2	1.5	0.4	0.3	0.2	—
Spent brewery yeast	—	7.0	0.4	0.3	0.04	0.03	—
Sweet potatoes	0.2	0.1	0.5	—	—	—	—
Tankage	7.0	1.5	3-10	—	—	—	—
Textile sludge	2.8	2.1	0.2	0.5	0.2	—	—
Wood ashes	0.0	2.0	6.0	20.0	1.0	—	—
Wood processing wastes	—	0.4	0.2	0.1	1.1	0.2	—
Tobacco stalks, leaves	3.7-4.0	0.5-0.6	4.5-6.0	—	—	—	—
Tobacco stems	2.5	0.9	7.0	—	—	—	—
Tomatoes, fruit, leaves	0.2-0.4	0.1	0.4	—	—	—	—

Note: Approximate values are given. Have materials analyzed for nutrient content before using.

* Adapted from J. P. Zublena, J. V. Baird, and J. P. Lilly, Extension Soil Science Specialists

North Carolina Cooperative Extension Service, Publication AG-439-18   June 1991.

(see http://ces.soil.ncsu.edu/soilscience/publications/Soilfacts/AG-439-18/)

Table D-2 . Nutrient Content of Manures (lb/unit wet basis) *

Type	TKN	P2O5	K2O	Ca	Mg	S
DAIRY						
Fresh (lb/ton)	10	5	8	4	2	1
Paved surface scraped(lb/ton)	10	6	9	5	2	2
Liquid manure (lb/1,000 lb)[1]	23	14	21	10	5	3
Lagoon liquid (lb/acre-inch)[2]	137	77	195	69	35	25
Anaerobic lagoon sludge (lb/acre-inch)[2]	15	22	81	2	4	4
BEEF						
Fresh (lb/ton)	12	7	9	5	2	2
Paved surface scraped (lb/ton)[1]	14	9	13	5	3	2
Unpaved feedlot (lb/ton)	26	16	20	14	6	5
Lagoon liquid (lb/acre-inch)2	83	77	129	24	19	—
Lagoon sludge (lb/1,000 lb)[1]	38	51	15	36	5	—
BROILER						
Fresh (lb/ton)	26	17	11	10	4	2
House litter (lb/ton)	72	78	46	41	8	15
Stockpiled litter (lb/ton)	36	80	34	54	8	12
DUCK						
Fresh (lb/ton)	28	23	17	—	—	—
House litter (lb/ton)	19	17	14	22	3	3
Stockpiled litter (lb/ton)	24	42	22	27	4	6
GOAT						
Fresh (lb/ton)	22	12	18	—	—	—
HORSE						
Fresh (lb/ton)	12	6	12	11	2	2
LAYERS						
Fresh (lb/ton)	26	22	11	41	4	4
Undercage paved (lb/ton)	28	31	20	43	6	7
Deep pit (lb/ton)	38	56	30	86	6	9
Liquid (lb/1,000 lb)[1]	62	59	37	35	7	8
Lagoon liquid (lb/acre-inch)[2]	179	46	26	62	57	52
Lagoon sludge (lb/1,000 lb)[1]	26	92	13	71	7	12
RABBIT						
Fresh (lb/ton)	24	23	13	19	4	2
SHEEP						
Fresh (lb/ton)	21	10	20	14	4	3
Unpaved (lb/ton)	14	11	19	24	7	6

Table E-2 .   Nutrient Content of Manures (lb/unit wet basis -continued)*

Type	TKN	P2O5	K2O	Ca	Mg	S
SWINE						
Fresh (lb/ton)	12	9	9	8	2	2
Surface scraped (lb/ton)	13	12	9	12	2	2
Liquid manure (lb/1,000 lb)1	31	22	17	9	3	5
Lagoon liquid (lb/acre-inch)[2]	136	53	133	25	8	10
Lagoon sludge (lb/1,000 lb)[1]	22	49	7	16	4	8
TURKEY						
Fresh (lb/ton)	27	25	12	27	2	—
House litter (lb/ton)	52	64	37	35	6	9
Stockpiled litter (lb/ton)	36	72	33	42	7	10

* J.P. Zublena, J.V. Baird, and J.P. Lilly, Extension Soil Science Specialists, N. Carolina Cooperative Extension Service, Publication AG-439-18, June 1991.

(see12/97. http://ces.soil.ncsu.edu/soilscience/publications/Soilfacts/AG-439-18/)

Notes: Approximate nutrient contents are given. Have materials analyzed for nutrient content before using. North Carolina mean waste analysis 1981 to 1990 supplied by J.C. Barker, NCSU Dept. Biological and Agricultural Engineering.

[1] Pounds per thousand pounds of manure liquid (slurry);

[2] Pounds per acre-inch. Estimated total lagoon liquid includes total liquid manure plus average annual lagoon surface rainfall surplus; does not account for seepage.

Appendix E

Directory of State Regulators

Entries followed by (B), refer to contacts for biosolids.
Entries followed by (SW), refer to contacts for solid waste.

Alabama
www.state.al.us

Water Div. (B)
334-271-7823

Division of Permits and Services (SW)
334-271-7714

Alabama Dept. Environmental Management
PO Box 301463
Montgomery, AL 36130-1463

Alaska
www.state.ak.us

Air and Water Quality Div. (B)
907-465-5010

Solid Waste Management (SW)
Div. Environmental Health
907-465-5162

Dept. Environmental Conservation
410 Willoughby Avenue, Suite 105
Juneau, AK 99801-1795

Arizona
www.adeq.state.az.us/environ/index.html

WaterQuality (B)
602-207-2306; 1-800-234-5677
Park Place, 500 N. Third Street
Phoenix, AZ 85004

Waste Programs (SW)
3033 North Central Avenue
Phoenix, AZ 85012
602-207-4117; 1-800-234-5677 x4117

Arkansas
www.adeq.state.ar.us

Water Div. (B)
501-682-0656

Solid Waste Div. (SW)
501-682-0600

Pollution Control and Ecology Dept.
8001 National Drive
PO Box 8913

Little Rock, AR 72219-8913

California
www.state.ca.us

Water Quality Div. (B)
Water Resources Control Board
901 P Street
Sacramento, CA 95814
916-657-0756

Integrated Waste Management Board
8800 Cal Center Drive (SW)
Sacramento, CA 95826
916-255-2200

Colorado
www.cdphe.state.co.us/environ.asp

Water Quality Control Div. (B)
303-692-3598

Environmental Office, Health Dept. (SW)
303-692-3000

Dept. of Health
4300 Cherry Creek Drive South
Denver, CO 80222-1530

Connecticut
www.dep.state.ct.us/ourenvir.htm

Water Management Bureau (B)
Permitting, Enforcement & Remediation
203-424-3705

Compost; Planning & Standards (SW)
Waste Management Bureau
203-424-3066

79 Elm Street
Hartford, CT 06106

District of Columbia
www.washingtondc.gov/agencies/

Water & Sewer Authority (B)
Environmental Regulation Administration
5000 Overlook Ave, SW
Washington, D.C. 20032
202-787-2000

Solid Waste (SW)
Dept. of Public Works
2000 14th Street NW, 6th Floor
Washington, D.C. 20009
202-673-6833

Delaware
www.dnrec.state.de.us/DNREC2000

Water Resources Div. (B)
302-739-4860

Air & Waste Management (SW)
302-739-3689

Natural Resources & Environmental Control
89 Kings Highway, PO Box 1401
Dover, DE 19903-1401

Florida
www.dep.state.fl.us/officsec/contact/

Water Facilities Div. (B)
Waste Management Div. (SW)
1-800-7414 DEP

Environmental Protection Dept.
3900 Commonwealth Boulevard
Tallahassee, FL 32399-3000

Georgia
www.ganet.org/dnr/environ/

Water Protection Bureau (B)
404-675-2692;1-888-373-5947

Solid Waste Management (SW)
404-675-2692

Environmental Protection Div.
205 Butler Street SE, Suite 1252
Atlanta, GA 30334

Hawaii
www.state.hi.us

Wastewater Bureau, 808-586-4185 (B)
Office Environmental Qual. Control (SW)
235 S. Beretania St., State Office

Commission on Water Resource Management,
PO Box 6212
Honolulu, HI 96813
808-587-0214

Idaho
www2.state.id.us/deq/waste/waste1.htm

Waste Management (B)
208-373-0298

Main State Office, 208-373-0502 (SW)

Dept. Environmental Quality
1410 N. Hilton St.
Boise, ID 83720-0036

Illinois
www.epa.state.il.us/

Water Pollution Control, 217-782-3397 (B)

Solid Waste, Bureau of Land (SW)
217-785-8604

Environmental Protection Agency
PO Box 19276
Springfield, IL 62794-9276

Indiana
www.state.in.us/idem/index/html

Water Management (B)
317-232-8470

Solid & Hazardous Waste (SW)
Office of Land Quality
317-233-5530

Dept. of Environmental Management
105 South Meridian Street
PO Box 6015
Indianapolis, IN 46206-6015

Iowa

www.state.is.us/

Water Quality & Wastewaer Bureau   (B)
515-281-8877

Solid Waste, Land Quality Bureau   (SW)
515-281-4968; compost: -8912

Environmental Protection
Dept. Natural Resources
Des Moines, IA 50319-0034

Kansas

www.kdhe.state.ks.us/

Permits & Compliance Unit   (B)
Bureau of Water, Building 283
785-296-5500

Waste Management   (SW)
Environmental Div., Building 740
785-296-1600

Dept. Health & Environment
Forbes Field, Topeka, KS 66620

Kentucky

www.nr.state.ky.us/nrepc/dep/dep2.htm

Water Div.   (B)
502-564-3410

Waste Management Div.   (SW)
502-564-6716

Natural Resources & Environmental
Protection Cabinet
14 Reilly Road
Frankfort, KY 40601

Louisiana

www.deq.state.la.us/

Environmental Compliance   (B)
1-888-763-5424

Solid Waste Permits   (SW)
225-765-0219

Dept. Environmental Quality
PO Box 82135
Baton Rough, La 70884-2135

Maine

janus.state.me.us/dep/home.htm

Bureau Land & Water Quality   (B)
Hazardous & Solid Waste   (SW)
207-287-7688; 1-800-452-1942

Dept. Environmental Protection
17 State House Section
Augusta, ME 04333

Maryland

www.mde.state.md.us/

Water & Wastewater Permit Program (B)
Water Management Administration
410-631-3375

Solid Waste Program   (SW)
Waste Management Administration
410-631-3318

Dept. Environment
2500 Broening Highway
Baltimore, MD 21224

Massachusetts

www.magnet.state.ma.us/dep/contact.htm

Sewage Div.   (B)
Water Resources Authority
617-788-4442

Charlestown Navy Yard
100 First Avenue
Boston, MA 02129

Dept. Environmental Protection   (SW)
Div. of Solid Waste Management
617-292-5974
1-800-462-0444

1 Winter Street, 4th Floor
Boston, MA 02108

Michigan
www.deq.state.mi.us/

Surface Water Quality Div.	(B)
Waste Management Div.	(SW)

Dept. Environmental Quality
PO Box 30473
Lansing, MI 48909-7973
517-373-1949

Minnesota
www.pca.state.mn.us/water

Water Quality Div.	(B)
Ground Water & Solid Waste Div.	(SW)

Pollution Control Agency
651-296-6300; 1-800-657-3864
TTY: 651-282-5332

520 Lafayette Road North
St. Paul, MN 55155-4194

Mississippi
www.deq.state.ms.us/

Groundwater & Solid Waste Div.

Surface Water Div.	(B)
Pollution Control Office	(SW)

601-961-5171, or –5036, or 5005

Environmental Quality Dept.
PO Box 20305
Jackson, MS 39289-1305

Missouri
www.dnr.state.mo.us/

Environmental Improvement Authority (B)
1-800-334-6946;
TDD: 1-800-379-2419

Solid Waste Management Program (SW)
Div. of Environmental Quality
573-751-5401

Dept. Natural Resources
PO Box 176
Jefferson City, MO 65102

Montana
www.deq.state.mt.us/

Water Quality Div.	(B)
Environmental Quality Dept.	(SW)

406-444-2544

PO Box 200901
Helena, MT 56920-0901

Nebraska
www.deq.state.ne.us

Water Quality	(B)
Air & Waste Management	(SW)

402-471-2186

Environmental Quality Dept.
1200 N Street, Suite 400
PO Box 98922
Lincoln, NE 68509-8922

Nevada
www.state.nv.us

Water Resources Div.   (B)
Conservation & Natural Resources Dept., 775-687-6972

Solid Waste, Bureau of Waste   (SW)
Management, Div. of Environmental
Protection, 775-687-4670,
1-800-597-5865

123 West Nye Lane
Carson City, NV 89706

New Hampshire
www.state.nh.us/

Water Supply & Pollution Control Div. (B)
603-271-2513; 603-271-2902

Waste Management Div.   (SW)
603-271-3644
1-800-273-9469
Dept. Environmental Services
Six Hazen Drive
Concord, NH 03301

New Jersey

www.state.nj.us/dep

Div. of Water Quality, Bureau of (B)
Pre-Treatment & Residuals Management;
609-633-3828

Solid Waste Regulation (SW)
609-633-1410

Dept. of Environmental Protection,
Trenton, NJ 08625

New Mexico

www.nmenv.state.nm.us/

Water & Waste Management Div. (B)
505-827-0187; 505-827-2918

Solid Waste Bureau (SW)
505-827-2775

Environment Dept.
1190 St. Francis Drive
PO Box 26110
Santa Fe, NM 87502

New York

www.dec.state.ny.us/

Water Div., Dept. Environmental Qual. (B)
518-457-6674

Solid & Hazardous Material (SW)
Environmental Quality Program
518-457-6934

50 Wolf Road
Albany, NY 12233

North Carolina

www.enr.state.nc.us

Water Resource Div. (B)
919-733-4064

Environment & Natural Resources
512 N. Salisbury St.
Raleigh, NC 27604

Solid Waste, 919-733-0692 (SW)
Dept. Environment & Natural Resources
401 Oberlin Rd. Suite 150
Raleigh, NC 27605

North Dakota

www.health.state.nd.us/ndhd/environ/wq

Water Quality Div. (B)
701-328-5210

Waste Management Div. (SW)
701-328-5166
Environmental Health
Health Dept.
1200 Missouri Avenue, PO Box 5520
Bismarck, ND 58502-5520

Ohio

www.epa.state.oh.us

Surface Water Div. (B)
614-644-2001

Drinking & Ground Water Div. (SW)
Solid & Infectious Waste Management
614-644-2909; 614-644-2621

Environmental Protection Agency
122 S. Front St., P.O. Box 1049
Columbus, OH 43216-1049

Oklahoma

www.deq.state.ok.us/

Water Quality Div. (B)
405-702-8100

Waste Management Div. (SW)
405-702-1000

Dept. of Environmental Quality
1000 NE 10th Street
Oklahoma City, OK 73117-1212

Oregon
www.deq.state.or.us

Wastewater Control (B)
Waste Management & Cleanup Div. (SW)

Dept. of Environmental Quality
811 SW Sixth Avenue
Portland, OR 97204
503-229-6442; 503-229-5913;
1-800-452-4011

Pennsylvania
www.dep.state.pa.us/dep/biosolids/biosolids.htm

Water Quality Management (B)
Dept. Environmental Protection
PO Box 8474
Harrisburg, PA 17105-8774
717-787-8184

Land Recycling & Waste Management
Dept. Environmental Protection (SW)
PO Box 8771
Harrisburg, PA 17105-8471
717-787-7816

Rhode Island
www.state.ri.us/dem/

Groundwater & Sewage Systems (B)
Water Quality Management
401-222-6820

Waste Management (SW)
401-222-2797

Dept. Environmental Management
235 Promenade Street
Providence, RI 02908

South Carolina
www.state.sc.us/dhec/

Water Quality Bureau (B)
Bureau Environmental Services
803-898-3432

Solid Waste Management Div. (SW)
Bureau of Land, Waste Management

Dept. Health and Environmental Control
803-896-4007

2600 Bull Street
Columbia, SC 29201

South Dakota
www.state.sd.us/denr/denr.html

Surface Water Quality (B)
605-773-3351

Office of Waste Management (SW)
605-773-3153

Dept. Environment & Natural Resources
523 East Capitol Avenue
Pierre, SD 57501-3181

Tennessee
www.state.tn.us/environment/wpc/

Water Pollution Control Div. (B)
615-532-0625

Solid Waste (SW)
615-532-0780

Dept. Environment & Conservation
401 Church Street, 21st Floor
Nashville, TN 37243-0435

Texas
www.tnrcc.state.tx.us/

Sludge & Transporter Review Unit (B)
512-239-4433

Waste Permits (SW)
512-239-2334

Natural Resource Conservation Commission
12100 Park 35 Circle, PO Box 13087
Austin, TX 78753

Utah
www.deq.state.ut.us

Water Quality Div.	(B)
801-538-6047	

Solid & Hazardous Waste Div. (SW)
801-538-6775

Dept. Environmental Quality
288 North 1460 West
Salt Lake City, UT 84116

Vermont
www.anr.state.vt.us

Wastewater Management Div. (B)
802-241-3739

Solid Waste Management Div. (SW)
802-241-3444

Vermont Agency of Natural Resources
State Complex, 103 South Main Street
Waterbury, VT 05671

Virginia
www.deq.state.va.us/info/direct.html

Water Div. (B)
804-762-4050

Waste Management Div.
804-762-4213

Dept. of Environmental Quality (SW)
629 East Main Street
PO Box 10009
Richmond, VA 23240-0009

Washington
www.ecy.wa.gov

Washington State Dept. of Ecology (SW)
PO Box 47600
Olympia, WA 98504-7600
360-407-6107

West Virginia
www.state.wv.us/directory/default.htm

Water Resources (B)
304-558-2107

Waste Management (SW)
304-558-2107

Environmental Protection Div.
10 McJunkin Road
Nitro, WV 25143

Wisconsin
www.dnr.state.wi.us/environment.html

Waste Water Management (B)
608-264-8954

Solid Waste Management Bureau (SW)
Dept. Natural Resources
Madison, WI 53707

Wyoming
www.deq.state.sy.us/index.htm

Water Quality Div. (B)
307-777-7075

Solid & Hazardous Waste Div. (SW)
307-777-7752

Environmental Quality Dept.
122 West 25th Street
Cheyenne, WY 82002

Appendix F

Glossary

Definitions of words used in this guidance document are listed here; the underlined words are defined elsewhere in this glossary.

AEROBIC

Living or active in the presence of oxygen. Used in this report to refer especially to microorganisms and/or decomposition of organic matter.

ANAEROBIC

Living or active in the absence of oxygen, e.g., anaerobic microorganisms.

ANIMAL (AND POULTRY) MANURE

Animal excreta, including bedding, feed and other by-products of animal feeding and housing operations.

BACTERIA

Single-celled microscopic organisms lacking chlorophyll. Some cause disease, and some do not. Some are involved in performing a variety of beneficial biological treatment processes including biological oxidation, solids digestion, nitrification, and denitrification.

BIOLOGICAL OXIDATION

The aerobic degradation of organic substances by microorganisms, ultimately resulting in the production of carbon dioxide, water, microbial cells, and intermediate byproducts.

BIOSOLIDS

The organic solids product of municipal wastewater treatment that can be beneficially utilized. Wastewater treatment solids that have received PSRP or PFRP treatment, or their equivalents, according to the Part 503 rule to acheive a Class A or Class B pathogen status.

The solids:liquid content of the product can vary:

- liquid biosolids 1-4% solids

- thickened liquid biosolids 4-12% solids

- dewatered biosolids 12-45% solids

- dried biosolids >50% solids (advanced alkaline stabilized, compost, thermally dried)

In general liquid biosolids and thickened liquids can be handled with a pump. Dewatered/dried biosolids are handled with a loader.

BOD (BIOCHEMICAL OXYGEN DEMAND)

The quantity of oxygen used in the biological and chemical oxidation (decomposition) of organic matter in a specified time, at a specified temperature (typically 5 days at 20°C), and under specified conditions. A standardized BOD test used in assessing the amount of organic matter in wastewater.

BUFFER

Around the perimeter of a storage or application area, a strip of land that is not intended to receive biosolids. The purpose of the buffer is to provide a protected zone around field boundaries, roads and sensitive areas, such as streams and wet soil areas.

BY-PRODUCT

A secondary or additional product; something produced in the course of treating or manufacturing the principal product.

CAKE

Dewatered biosolids with a solids concentration high enough (>12%) to permit handling as a solid material. (Note: some dewatering agents might still cause slumping even with solids contents higher than 12%).

CATION EXCHANGE CAPACITY (CEC)

A measure of the soil's capacity to attract and retain plant nutrients that occur in positively charged ionic form. CEC is a focus of interest because fertilizers supply positively charged cationic plant nutrients, which are attracted to negatively charged anionic soil particles, including soil organic matter. Organically amended soils typically have a higher CEC, i.e., a higher capacity for attracting and retaining plant nutrients, than unamended or low organic soils.

CFU (COLONY FORMING UNITS)

A term used to enumerate microbes in a sample and based on the fact that the visible cluster (colony) of microbes that appears on nutrient agar medium in a petri dish can develop from a single or group of microbial cells.

COMPOSTING

The accelerated decomposition of organic matter by microorganisms, which is accompanied by temperature increases above ambient; for biosolids, composting is typically a managed aerobic process.

CONSOLIDATED (BIOSOLIDS)

A desirable characteristic of biosolids that allows the material to be stacked and remain non-flowing when stored.

CRITICAL CONTROL POINT

A location, event or process point at which specific monitoring and responsive management practices should be applied.

DENITRIFICATION

The conversion of nitrogen compounds to nitrogen gas or nitrous oxide by microorganisms in the absence of oxygen.

DEWATERED BIOSOLIDS

The solid residue (12% total solids by weight or greater) remaining after removal of water from a liquid biosolids by draining, pressing, filtering or centrifuging. Dewatering is distinguished from thickening in that dewatered biosolids may be transported by solids handling procedures.

DIGESTION

Decomposition of organic matter by microorganisms with consequent volume reduction. Anaerobic digestion produces methane and carbon dioxide, whereas aerobic digestion produces carbon dioxide and water.

EQ BIOSOLIDS

Exceptional Quality biosolids, meets Class A pathogen reduction, and Vector Attraction Reduction standards 1- 8, and Part 503, Table 3 high quality pollutant concentration standards.

EUTROPHICATION

A natural or artificial process of nutrient enrichment by which a water body becomes highly turbid, depleted in oxygen, and overgrown with undesirable algal blooms.

FECAL COLIFORM

The type of coliform bacteria present in virtually all fecal material produced by mammals. Since the fecal coliforms may not be pathogens, they indicate the potential presence of human disease organisms. See indicator organisms.

FECAL STREPTOCOCCUS

A member of a group of gram-positive bacteria known as *Enterococci*, previously classified as a subgroup of *Streptococcus*. They are found in feces of humans, animals, and insects on plants often not in association with fecal contamination. See indicator organisms.

FIELD STORAGE

Temporary or seasonal storage area, usually located at the application site, which holds biosolids destined for use on designated fields. State regulations may or may not make distinctions between staging, stockpiling, or field storage. In addition, the time limits for the same material to be stored continuously on site before it must be land applied range from 24 hours to two years.

FILTER PRESS

Equipment used near the end of the solids production process at a wastewater treatment facility to remove liquid from biosolids and produce a semi-solid <u>cake</u>.

GENERATOR

Person or organization producing or preparing the <u>biosolids</u> by treatment of <u>wastewater</u> solids. Also, a person or organization who changes the biosolids characteristics either through treatment, mixing or any other process.

GOOD MANAGEMENT PRACTICES

Schedules of activities, operation and maintenance procedures (including practices to control odor, site runoff, spillage, leaks, or drainage), prohibitions, and other management practices found to be highly effective and practicable in the safe, community-friendly use of <u>biosolids</u> and in preventing or reducing discharge of pollutants to waters of the United States.

HELMINTH AND HELMINTH OVA

Parasitic worms, e.g., roundworms, tapeworms, *Ascaris, Necator, Taenia,* and *Trichuris,* and ova (eggs) of these worms. Helminth ova are quite resistant to chlorination, and can be passed out in the feces of infected humans and organisms and ingested with food or water. One helminth ovum is capable of hatching and growing when ingested.

HYDRAULIC LOADING RATES

Amount of water or <u>liquid</u> <u>biosolids</u> applied to a given treatment process and expressed as volume per unit time, or volume per unit time per surface area.

INDICATOR ORGANISMS

Microorganisms, such as <u>fecal coliforms</u> and <u>fecal streptococci</u> (enterococci), used as surrogates for bacterial pathogens when testing biosolids, manure, compost, leachate and water samples. Tests for the presence of the surrogates are used because they are relatively easy, rapid, and inexpensive compared to those required for <u>pathogens</u>, such as <u>salmonella</u> bacteria.

INFILTRATION

The rate at which water enters the soil surface, expressed in inches per hour, influenced by both <u>permeability</u> and moisture content of the soil.

LAGOON

A reservoir or pond built to contain water, sediment and/or manure usually containing 4% to 12% solids until they can be removed for application to land.

LAND APPLICATION

The spreading or spraying of <u>biosolids</u> onto the surface of land, the direct injection of biosolids below the soil surface, or the incorporation into the surface layer of soil; also applies to manure and other organic residuals.

LEACHATE

Liquid which has come into contact with or percolated through materials being stockpiled or stored; contains dissolved or suspended particles and <u>nutrients</u>.

LIQUID BIOSOLIDS OR MANURE

Biosolids or animal manure containing sufficient water (ordinarily more than 88 percent) to permit flow by gravity or pumping.

MERCAPTANS

A group of volatile chemical compounds, that are one of the breakdown products of sulfur-containing proteins. Noted for their disagreeable odor.

MICROORGANISM

Bacteria, fungi (molds, yeasts), protozoans, helminths, and viruses. The terms *microbe* and *microbial* are also used to refer to microorganisms, some of which cause disease, and others are beneficial. Parasite and parasitic refer to infectious protozoans and helminths. Microorganisms are ubiquitous, possess extremely high growth rates, and have the ability to degrade all naturally occurring organic compounds, including those in water and wastewater. They use organic matter for food.

MINERALIZATION

The process by which elements combined in organic form in living or dead organisms are eventually reconverted into inorganic forms to be made available for a new cycle of growth. The mineralization of organic compounds occurs through oxidation and metabolism by living microorganisms.

MITIGATION

The act or state of reducing the severity, intensity, or harshness of something; to alleviate; to diminish; to lessen; as, to mitigate heat, cold, or odor.

MPN (MOST PROBABLE NUMBER)

A statistically approximation of the number of microorganisms per unit volume or mass of sample. Often used to report the number of coliforms per 100 ml wastewater or water, but applicable to other microbial groups as well.

NITRIFICATION

The biochemical oxidation of ammonia nitrogen to nitrate nitrogen, which is readily used by plants and microorganisms as a nutrient.

NONPOINT SOURCE

Any source, other than a point source, discharging pollutants into air or water.

NONPOINT SOURCE POLLUTION

Man-made or man-induced alteration of the chemical, physical, biological, or radiological integrity of water or air, originating from any source other than a point source.

NUTRIENT

Any substance that is assimilated by organisms and promotes growth; generally applied to nitrogen and phosphorus in wastewater, but also other essential trace elements or organic compounds that microorganisms, plants, or animals use for their growth.

NUTRIENT MANAGEMENT PLAN

A series of good management practices aimed at reducing agricultural nonpoint source pollution by balancing nutrient inputs with crop nutrient requirements. A plan includes soil testing, analysis of organic nutrient sources such as biosolids, compost, or animal manure, utilization of organic sources based on their nutrient content, estimation of realistic yield goals, nutrient recommendations based on soil test levels and yield goals, and optimal timing and method of nutrient applications.

ODOR CHARACTER

The sensory quality of an odorant, defined by one or more descriptors, such as fecal (like manure), sweet, fishy, hay, woody resinous, musty, earthy, see Atlas of Odor Character Profiles, 1985.

ODOR DILUTIONS TO THRESHOLD or D/T

Dimensionless unit expressing the strength of an odor. An odor requiring 500 binary (2-fold) dilutions to reach the detection threshold has a D/T of 500. An odor with a D/T of 500 would be stronger than an odor with a D/T of 20.

ODOR INTENSITY

A measure of the perceived strength of an odor. This is determined by comparing the odorous sample with "standard" odors comprised of various concentrations of n-butanol in odor-free air.

ODOR PERVASIVENESS

Persistence of an odor; how noticeable an odorant is as it's concentration changes; determined by serially diluting the odor and measuring intensity at each dilution.

ODOR THRESHOLD

Detection - The minimum concentration of an odorant that, on average can be detected in odor-free air.

Recognition - The minimum concentration of an odorant that, on average, a person can distinguish by its definite character in a diluted sample.

OFF-SITE STORAGE

Storage of biosolids at locations away from the wastewater treatment plant or from the point of generation. Several terms encompass various types of storage: Staging, Stockpiling, Field Storage, and Storage facility.

OVERLAND FLOW

Refers to the free movement of water over the ground surface.

PATHOGEN

Disease-causing organism, including certain bacteria, fungi, helminths, protozoans, or viruses.

PERMEABILITY

The rate of liquid movement through a unit cross section of saturated soil in unit time; commonly expressed in inches per hour.

PFRP , PSRP

See <u>Process to Further Reduce Pathogens</u>, or <u>Process to Significantly Reduce Pathogens</u>

pH

A measure used to indicate the degree of acidity or alkalinity of a substance. The pH is expressed as the $\log_{10}$ of the reciprocal of the actual hydrogen ion concentration. The pH ranges from 0-14, where 0 is the most acidic , 14 is the most alkaline, and 7 is neutral.

PHYTOTOXIN

Any substance having a toxic or poisonous effect on plant growth. Immature or anaerobic compost can contain volatile fatty acids that are phytotoxic to plants. Soluble salts can also be phytotoxic in addition to toxic heavy metals and toxic organic compounds.

POINT SOURCE

Any discernable, confined, or discrete conveyance from which pollutants are or may be discharged, including, but not limited to, any pipe, ditch, channel, tunnel, conduit, well, stack, container, rolling stock, concentrated animal feeding operation, or vessel or other floating craft.

POLYMER

A compound composed of repeating subunits used to aid in flocculating suspended particulates in wastewater into large clusters. This flocculation aids solids removal, and enhances the removal of water from biosolids during dewatering processes.

PROCESS TO FURTHER REDUCE PATHOGENS (PFRP)

The process management protocol prescribed in U.S. EPA Part 503 used to achieve Class A <u>biosolids</u> in which pathogens are reduced to undetectable levels. Composting, advanced alkaline stabilization, chemical fixation, drying or heat treatment, are some of the processes that can be used to meet Part 503 requirements for Class A.

PROCESS TO SIGNIFICANTLY REDUCE PATHOGENS (PSRP)

The process management protocol prescribed in U.S. EPA Part 502 used to achieve Class B <u>biosolids</u> in which pathogen numbers are significantly reduced, but are still present. Additional restrictions on the use and placement of Class B biosolids ensure a level of safety equivalent to Class A. <u>Aerobic</u> and <u>anaerobic</u> digestion, air drying and lime stabilization are types of processes used to meet the Class B pathogen density limit of less than 2,000,000 fecal coliforms/gram dry weight of total solids.

PROTOZOA

Single-celled, <u>microorganism</u>s many species of which can infect man and cause disease. The infective forms are passed as cysts or oocysts in the feces of man and animals and accumulate in flocculated solids; they are quite resistant to disinfection processes, such as chlorination, that eliminate most <u>bacteria</u>, but are susceptible to destruction by drying.

RETENTION TIME

The period of time <u>wastewater</u> or <u>biosolids</u> takes to pass through a particular part of a treatment process, calculated by dividing the volume of processing unit by the volume of material flowing per unit time.

RISK, POTENTIAL

Refers to a description of the pathways and considerations involved in the occurrence of an event (or series of events) that may result in an adverse health or environmental effect.

RISK ASSESSMENT

A quantitative measure of the probability of the occurrence of an adverse health or environmental effect. Involves a multi-step process that includes hazard identification, exposure assessment, dose-response evaluation, and risk characterization. The latter combines this information so that risk is calculated:

Risk = Hazard x Exposure

RUNOFF

That part of the precipitation that runs off the surface of a drainage area when it is not absorbed by the soil.

SALMONELLA

Rod-shaped <u>bacteria</u> of the genus *Salmonella*, many of which are <u>pathogenic</u>, causing food poisoning, typhoid, and paratyphoid fever in human beings, or causing other infectious diseases in warm-blooded animals, and can cause allergic reactions in susceptible humans, and sickness, including severe diarrhea with discharge of blood.

SEPTAGE

Domestic sewage (liquid and solids) removed from septic tanks, cesspools, portable toilets, and marine sanitation devices; not commercial or industrial wastewater.

SEWAGE, DOMESTIC

Residual liquids and solids from households conveyed in municipal wastewater sewers; distinguished from wastewater carried in dedicated industrial sewers. See <u>Wastewater</u>.

SLUMPING

Failure of a stockpile to retain a consolidated shape usually due to insufficient dewatering of the <u>biosolids</u>. Slumping may result in biosolids movement beyond the boundaries of a designated stockpile area or create handling difficulties when the materials are scooped up and loaded into spreaders.

SOLIDS

In water and <u>wastewater</u> treatment, any dissolved, suspended, or <u>volatile</u> substance contained in or removed from water or wastewater.

STABILITY

The characteristics of a material that contribute to its resistance to decomposition by microbes, and to generation of odorous metabolites. The relevant characteristics include the degree of organic matter decomposition, nutrient, moisture and salts content, pH, and temperature. For <u>biosolids</u>, <u>compost</u>, or animal manure, stability is a general term used to describe the quality of the material taking in to account its origin, processing, and intended use.

STAGING

The concurrent delivery and application of <u>biosolids</u>, allowing for the transfer of biosolids from transport vehicles to land application equipment. Dewatered materials may be off-loaded from delivery vehicles to temporary stockpiles to facilitate the loading of spreading equipment.

STOCKPILING

Holding of <u>biosolids</u> at an active field site long enough to accumulate sufficient material to complete the field application.

STORAGE

Placement of Class A or B biosolids in designated locations (other than the WWTP) until material is land applied; referred to as field storage. See also Off-Site Storage.

STORAGE FACILITY

An area of land or constructed facilities committed to hold biosolids until the material may be land applied at on- or off-site locations; may be used to store <u>biosolids</u> for up to two years. However, most are managed so that biosolids come and go on a shorter cycle based on weather conditions, crop rotations and land availability, equipment availability, or to accumulate sufficient material for efficient spreading operations.

THRESHOLD ODOR

See Odor Threshold

TURBULENCE

Irregular atmospheric motion especially characterized by up and down currents. Increasing turbulence results in dilution of odors.

VAR

Abbreviation for Vector Attraction Reduction (see Appendix C, Table C-3).

VECTOR

An agent such as an insect, bird, animal, that is capable of transporting <u>pathogens</u>.

VIRUS

A microscopic, non-filterable biological unit, technically not living, but capable of reproduction inside cells of other living organisms, including bacteria, protozoa, plants, and animals.

VOLATILE COMPOUND

A substance that vaporizes at ambient temperature. Above average heat can increase the volatilization (vaporization) rate and amount of many volatile substances.

WWTP

Abbreviation for wastewater treatment plant

MUNICIPAL WASTEWATER

Household and commercial water discharged into municipal sewer pipes; contains mainly human excreta and used water. Distinguished from solely industrial wastewater.

WASTEWATER TREATMENT

The processes commonly used to render water safe for discharge into a U.S. waterway: 1) Preliminary treatment includes removal of screenings, grit, grease, and floating solids; 2) Primary treatment includes removal of readily settleable organic solids; 50-60% suspended solids are typically removed along with 25-40% BOD; 3) Secondary treatment involves use of biological processes along with settling; 85-90% of BOD and suspended solids are removed during secondary treatment; 3) Tertiary treatment involves the use of additional biological, physical, or chemical processes to remove more of the remaining nutrients and suspended solids.

www.ingramcontent.com/pod-product-compliance
Lightning Source LLC
Chambersburg PA
CBHW080640180526
45168CB00008B/3242